THE SAVAGE GARDEN

REVISED

Cultivating Carnivorous Plants

PETER D'AMATO

TEN SPEED PRESS
Berkeley

I dedicate this book to my biz partners,
Damon Collingsworth and Mickey Urdea, and
to our employees at California Carnivores,
Norma McFaddan, Liz Brown, Axel Bostrom,
and Tom Kahl—all of us slaves of
flesh-eating plants.

Published in the United States
by Ten Speed Press, an imprint of
the Crown Publishing Group, a division
of Random House, Inc., New York.
www.crownpublishing.com
www.tenspeed.com
Ten Speed Press and the Ten Speed Press colophon are registered
trademarks of Random House, Inc.

A previous edition of this work was published in the United States
in paperback by Ten Speed Press, Berkeley, in 1998.

All photography and illustration credits appear on page 362

Library of Congress Cataloging-in-Publication Data
D'Amato, Peter.
The savage garden / Peter D'Amato. — Rev.
p. cm.
1. Carnivorous plants. I. Title.
SB432.7.D35 2013
635.9'3375—dc23

2012047904

Trade Paperback: ISBN 978-1-60774-410-8
eBook: ISBN 978-1-60774-411-5

Printed in China

Design by Chloe Rawlins
Front cover and spine photographs by Patrick Hollingsworth
Back cover photograph (upper) by Jonathan Chester/Extreme Images, Inc.
Back cover photograph (lower) by Damon Collingsworth

10 9 8 7 6 5 4 3 2 1

Revised Edition

THE SAVAGE GA

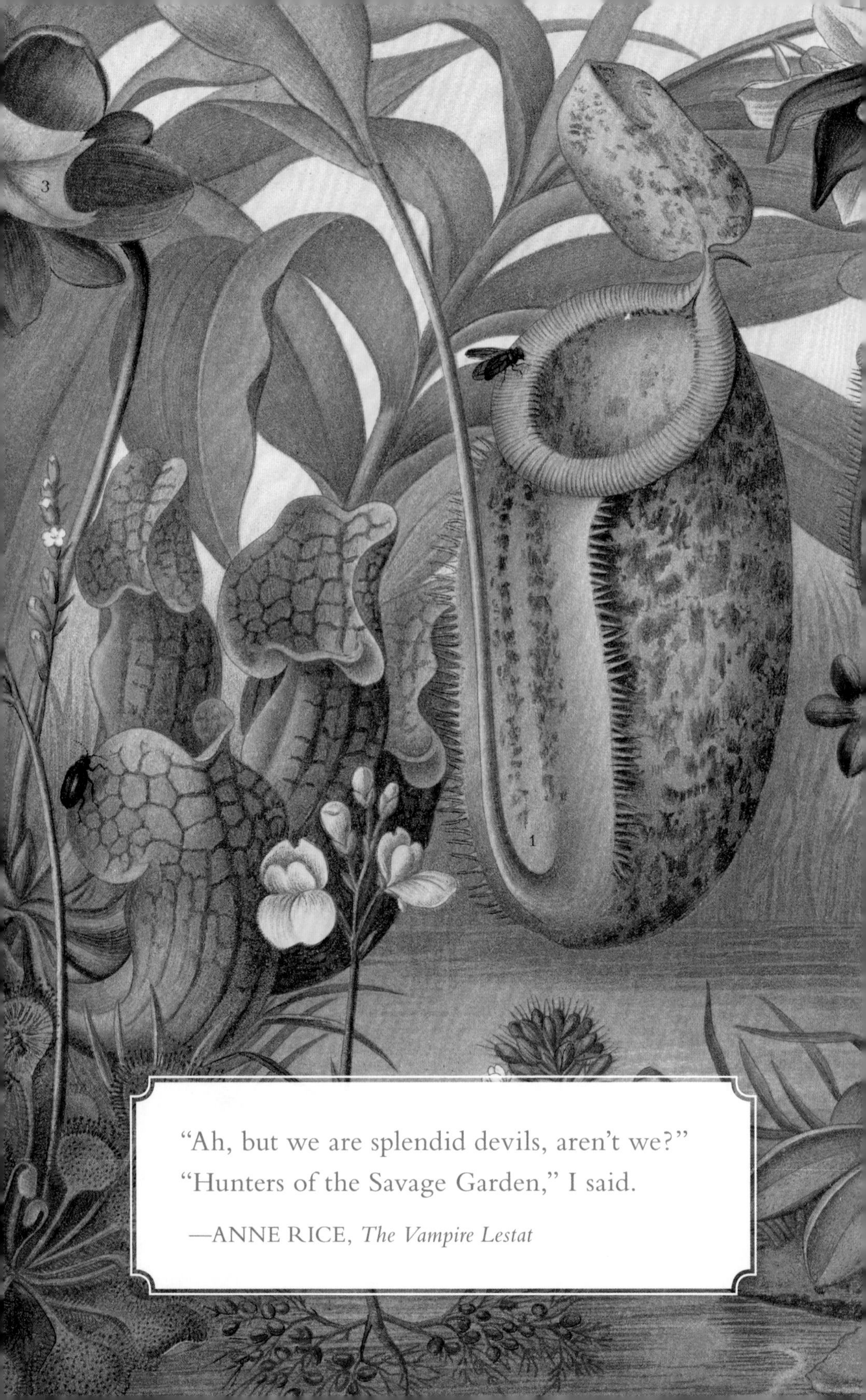

"Ah, but we are splendid devils, aren't we?"
"Hunters of the Savage Garden," I said.

—ANNE RICE, *The Vampire Lestat*

CONTENTS

PREFACE

I find it ironic that this introduction to the revised edition of *The Savage Garden* is the final section of the book I am completing, only one week after returning from the 2012 International Carnivorous Plant Society (ICPS) conference hosted by the New England Carnivorous Plant Society. I had a wonderful time at the conference, and in a presentation I gave, I briefly told the audience some details of this new edition. The irony is that at the very first world conference of the ICPS, held at the Atlanta Botanical Garden in 1997, I gave a slide show preview of the first edition of the book. Little did I truly comprehend at the time the potential impact of the book. It won the American Horticultural Society's book award in 1999 and became a best seller in the genre, and only two weekends ago in New England I autographed dozens of copies for attendees of the conference. Nearly all of the books were well worn, well read, many with colorful bits of paper bookmarking sections of the book important to the reader—well, I was humbled and pleased.

The Savage Garden was not just a coffee table book of pretty pictures, but a resource that people actually read for the helpful information it contained. It was the type of book I wish had existed when I was a kid in the late 1960s, when carnivorous plants first took hold of my life, yet information on them—especially how to grow them—was sorely lacking.

There have been many ICPS conferences since 1997 all over the world, from Japan to Australia, from Europe to North America. The one this past month in New England was startling not only for the amazing show of fantastically beautiful carnivorous plants on display and for sale, but also for the delightful variety of people attending. Adults and kids, men and women, boys and girls—things have certainly changed since I was a teenager—when you could count on one hand the serious growers of carnivorous plants in the United States. Since then, thousands of people all over the world have found the hobby an intoxicating experience. Carnivorous plants are educational, sometimes challenging, utterly and strangely beautiful, and a lot of fun to grow! Plus, unlike other plants—*they don't just sit there!*

So how does this revision differ from the first edition? A lot has happened in fifteen years.

Most of the cultivation information has remained primarily intact, but I have tweaked and fine-tuned many details learned over the years. Much of this has come from customers of my nursery, California Carnivores, who have done things unimaginable a decade and a half ago. Other changes came from continual experiments at our nursery, resulting in adjustments to soil recipes, or discovering the cold and freeze tolerance of various subtropical plants. The results were often surprising.

Perhaps the greatest changes have occurred in the identification of many of the plants and the astounding numbers of new species discovered and brought into cultivation in the past fifteen years.

When I first wrote *The Savage Garden*, for example, most of the varieties of *Sarracenia*, the American pitcher plants, had nicknames only, such as "red tube" or "coppertop" or "okee giant." All of these plants, and many new ones discovered and cultivated since, now have true names (see Registered Cultivars, page 357).

The vast number of new species that have been discovered and identified is truly amazing. When I first wrote the book, there were five known species of South American sun pitchers (*Heliamphora*). Now, the count is approaching twenty-five! The butterworts (*Pinguicula*) have jumped from seventy species to more than one hundred. And when I first wrote the book there were around eighty species of tropical pitcher plants (*Nepenthes*) known—the count is now approaching one hundred fifty—with many more expected to be discovered as modern-day enthusiasts climb the countless unexplored mountains from the Philippines to New Guinea. More carnivorous plants have been discovered in the last decade than at any time in the past.

I have also updated some technological advances, like the invention of polycarbonate greenhouse glazing, which is far superior to fiberglass and glass. Grow light technology has also improved greatly, with far superior artificial lights on the market. We have discovered a new fertilizer that doesn't kill live sphagnum moss, and many new insecticides—some completely natural—that can control bothersome pests with no harm to plants or people.

I have also added the metric system to the revision—a tedious effort that nearly drove me insane. Please keep in mind these systems of measurements are rough and approximate!

I want to thank my business partner, Damon Collingsworth, for taking on the task of coordinating all the photographs for the book. He took dozens of new photos at our nursery and solicited many others from our friends both near and far. As just one example, the photography of plant autopsies, taken by California Carnivores volunteer Patrick Hollingsworth, is unsurpassed. And the astounding images lent to us by naturalists like Ch'ien Lee, Fernando Rivadavia, Stewart McPherson, Barry Rice, and Andreas Fleischman are certainly compelling and hypnotic.

I must end this preface to a book on a topic that has brought so much joy to so many people around the world with a message of sadness and pain.

After the recent conference in New England, many of the attendees went on field trips to see carnivorous plants in the wild. They saw local bogs in New England as well as sites in New Jersey, North Carolina, and the Gulf Coast of southeastern North America, where more species of carnivorous plant genera exist than anywhere else in the world. The excitement of seeing several habitats in protected, pristine beauty was unfortunately overshadowed by one heartbreaking fact: these astounding forms of life are disappearing from our planet at a rate that is beyond alarming.

As a few examples, I went to southern New Jersey to see a couple of my favorite sites I knew as a kid growing up on the Jersey shore. Both of these habitats had originally been bogs in the 1800s, but the slow-moving streams had been dammed to create lake reservoirs for drinking water by the early 1900s. Still, there were vast numbers of surviving plants along the edges of the lakes when I was a kid. Apparently, however, the dams were rebuilt in the past decade, and when I saw them a couple of weeks ago, the enormous quantity of sphagnum moss, pitcher plants, and sundews I remembered were virtually all gone, drowned by the higher lake levels the new dams created. And this was in the Pinelands National Reserve, which "protects" an amazing 22 percent of the small and densely populated state of New Jersey.

Around Wilmington, North Carolina, we saw some very nice, small sites on Nature Conservancy property that is well managed with controlled burns and protected by a locked gate across the road. However, beyond that, we were left having to hunt for plants along dismal drainage ditches along the sides of roads where occasional patches of Venus flytraps, sundews, and pitcher plants barely survived among broken bottles and discarded fast-food containers.

In fact, the most Venus flytraps we saw were in the back yard lawn of a kindly old gentleman who loved and protected the pitcher plants, flytraps, and sundews that survived on the edge of his lakefront property. Originally the area was a vast bog of carnivorous and other rare plants. A few decades ago the creek that drained the bog was dammed, a large lake flooded the bog, and lovely estates were built along the lake. The few plants that survived now grow in grassy lawns, if they are not removed by the homeowner.

A simple ditch dug one or two feet (0.3–0.6 m) deep along the edge of a wet carnivorous plant habitat can drain vast areas, killing off all the plants that rely on wet soils. As Mark Todd, our host in Wilmington and member of the North American *Sarracenia* Conservancy (http://nasarracenia.org), told us, "There are just fewer and fewer habitats where the plants are able to grow." And sadly, this is occurring all around the world.

Thankfully, many readers of *The Savage Garden* first encounter the book to find out how to grow a Venus flytrap or pitcher plant they've rescued from a local nursery. If this is the case for you, I hope that your initial interest will grow and evolve, leading you to support the various organizations that are trying to save these wonders of nature from extinction.

INTRODUCTION: WHAT ARE CARNIVOROUS PLANTS?

"The Venus flytrap, a devouring organism, aptly named for the goddess of love."
—Tennessee Williams, *Suddenly Last Summer*

I was a kid growing up in the sixties when I had my first love affair with a Venus flytrap.

It was an advertisement in *Famous Monsters* magazine that first seduced me. The ad shouted something about the plant eating hamburger, and next to it was a fuzzy picture of Charles Darwin. As I already had pet turtles and a South American alligator trained to eat hamburger from a spoon, I convinced my mother that we could afford another mouth to feed and promptly had her write out a check.

The plants finally arrived in a Styrofoam pot wrapped in plastic. The pot was filled with dry peat moss and three or four "bulbs" with all of their leaves cut off. I followed the directions, but nothing spectacular occurred. A few semideveloped leaves came up, but soon all of the plants turned black.

It was my first experience with unrequited love. It never crossed my mind that the dim corner of my bedroom next to the heater in the month of December had anything to do with the plants' demise. My first plant, a parlor palm, was happily thriving there. Venus flytraps, I thought, must come from some dark, steamy, tropical jungle—didn't they?

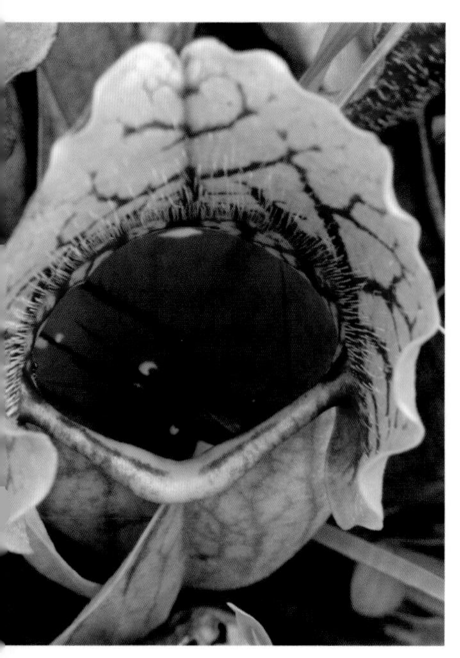

"I saw struggling insects in the wells of water each leaf held."

I was surprised when the following spring a friend and fellow student whispered to me, "I know where Venus flycatchers grow." I had just raised my hand in science class and volunteered to do a report on the Venus flytrap. Hope springs eternal; I figured that doing a school report was one way I could convince my mom to write out another check, and I could give Venus flytraps a second try. Since we lived on the seashore of southern New Jersey, I found my friend's statement rather hard to believe.

He took me to the boggy edge of a small lake right in the middle of town. The ground was covered in a billowy, spongy green moss that my friend called *sphagnum*. The moss hugged the bases of southern white cedars that grew in the shallow, tea-colored water. It was a beautiful sight.

"There they are," my friend said, pointing. I looked in awe at the strangest plants I had ever seen. Half buried in the moss were rosette clumps of deep purple, hollow leaves, with spiny collars and strange reddish flowers rising from the center.

"These aren't Venus flytraps," I said, but I was hardly disappointed.

When my friend assured me they still ate bugs, I peered into one of the hollow, leathery leaves, and sure enough I saw insects struggling in the wells of water that each leaf held.

"Look at this," my friend said. He plucked something from the moss and held it up in his fingers. It was an image that would forever be imprinted on my brain. A ray of sunlight broke through the cedars, shining directly on what he held in his hand. It was a small, circular green leaf covered with hundreds of red tentacles like a pincushion, each ending in a tiny drop of dew. Every drop caught the light of the sun, and they sparkled and glittered like jewels. These small plants were dotted with numerous dead and struggling insects, their circular leaves sometimes clenched like tight fists, with wings and antennae sticking out and twitching. I looked around in awe, for it was an unforgettable image: tea-colored water, grayish trunks of cedars, and spongy mounds of reddish green islands with strange plants that looked like they came from outer space!

"It was an image that would forever be imprinted upon my brain."

My friend and I dug up some of the weird plants and took them to school the following week. Even our teachers were mystified. But soon I was led to the library and found a couple of books and *National Geographic* articles that satisfied my curiosity. What we had found growing on the swampy edge of the lake were purple pitcher plants and sundews—carnivorous plants not unlike the famous Venus flytrap! I was also surprised to learn that the pine barrens of southern New Jersey were practically teeming with flesh-eating plants, and that the flytrap was native only to the Carolinas, a mere day's drive south of where I lived. I was almost dumbfounded to discover that North America has probably the widest variety of carnivorous plants in the world: pitcher plants, sundews, butterworts, cobra plants, bladderworts, and Venus flytraps all grow here! I wouldn't have to fly to Madagascar after all.

For me, it was the beginning of a mind-boggling adventure that would change my life.

Our general impression is that plants are fairly passive forms of life. Insects and animals eat them. We chop down trees to build houses, shred cabbage for coleslaw, and decorate our homes with their sex organs, which we call flowers. We eat their fruit, pull "weeds," and make medicine out of their sap. We bleed trees for maple syrup and burn them in fireplaces. We bake them, boil them, and sauté or stir-fry them. We even smoke them.

Plants can't scream and run away, but some of them do fight back to an extent. Mushrooms can kill you, and poison ivy can make you itch. Many plants defend themselves with needles and toxins or bitter tastes and bad smells.

Typically we are not afraid of plants, but humans love to project their own fears onto other life forms. That we can do this with plants, seemingly the most passive and unfrightening of life forms on earth, is obvious by just examining some of our more popular horror movies. *The Thing* featured an alien humanoid plant that fed on human blood. In *Day of the Triffids*, walking plants were stinging humans to death in their effort to take over the world. *Invasion of the Body Snatchers* had plant "pods" duplicating human beings and taking over their minds and bodies. In *Little Shop of Horrors*, a talking plant with a sense of humor swallows people whole. The fact that these four famous horror movies were made—and remade—reflects the unconscious fear that we all have of pretty, pulpy, passive plants. Perhaps deep within our brains, tiny neurons still fire off flashbacks of ancient, inherited memories—horrible memories of the days when our ancestors had good reason to fear plants!

You might smirk and shake your head, but this primal fear may not be quite as far-fetched as you think. Many years ago at California Carnivores, the carnivorous plant nursery I own, where we have on public display hundreds of varieties of flesh-eating plants, a well-to-do couple came in and were marveling at our collection of tropical pitcher plants, the *Nepenthes*. They told me with glee how they had recently returned from Malaysia, where they saw magnificent *Nepenthes* at a botanical garden. "We arrived early," the wife told me, "and we waited in line at the front gate. I peered through the fence, and saw these huge pitcher plants hanging in the trees. To my shock, an attendant was pulling tiny baby monkeys out of the traps! Most were alive, and scampered away. The dead ones he dropped in a pail." Her husband added that when they later caught up with the attendant and asked him about what he was doing, the embarrassed attendant explained that dead monkeys in the pitcher plants were upsetting to the tourists. While I have never been able to document this story, and I suspect the prey were more akin to rodents, such tales have persisted for a very long time—a mythology that may have a basis in fact.

Thus the following equation is not necessarily true: Plant eats monkey. Monkeys are primates. Humans are primates. Plant eats humans.

But surprisingly, there was a brief time when some people truly believed in the Man-Eating Tree of Madagascar. It was in 1878 that Carle Liche wrote an article claiming he had witnessed a sacrifice of a young maiden to such a tree by natives of the island. Since he offered grisly details and was published in scientific and popular magazines, the report was widely believed to be true. It was not.

During the same period Charles Darwin, among many others, studied and reported on the amazing carnivorous habits of many plants both familiar to them and being newly discovered around the world. Science was exploding in popularity in the nineteenth century, and explorations were uncovering many strange and exotic forms of life. Even plants long familiar to Europeans, like the sundews and butterworts common to local bogs and moors, were suddenly suspect.

This suspicion over which plants are or are not carnivorous is a matter that is not quite settled even in the early twenty-first century. "Carnivorous" means "flesh-eating." An older term, used by Darwin and sometimes used even today, was "insectivorous," or insect-eating. The latter term is not quite accurate and is rather limiting, for even though the vast majority of prey eaten by carnivorous plants are insects, this group of plants also consumes spiders, sow bugs, worms, tadpoles, frogs, lizards, and even rats, although admittedly the capture of larger animals such as mammals is a very rare event.

A typical "normal" plant works in the following way: the roots in the soil absorb water, including minerals. The leaves in the air absorb carbon dioxide. Through the complex process of photosynthesis, chlorophyll in the leaves uses the energy of sunlight to transform the carbon dioxide and minerals into carbohydrates and other organic compounds, which give the plant energy to grow.

But what if the soil a plant lives in is low in minerals—particularly nitrogen, phosphorous, or potassium, which are vital to the plant's health?

Most carnivorous plants grow in mineral-deficient soils. More often than not, these soils are very wet. The water moving through the ground carries away most of the much-needed minerals. Even nitrogen, returned to the soil by the slow decay of older, dying leaves, does not remain for long. A plant living in such an environment might be able to survive year after year, but it wouldn't be able to manufacture the energy needed to produce flowers, seed, or offshoots.

Carnivorous plants have an answer to that survival dilemma. All around them are little moving packets of minerals and nutrients, like vitamin pills with legs and wings. We call them animals. All the plant has to do is catch them and somehow absorb through their leaves what they would normally take up through their roots. The development of leaves for such a purpose is what makes carnivorous plants so bizarre and beautiful.

Even "normal" plants can absorb minerals through their leaves. You can spray your rosebush with a watery solution of Miracle-Gro fertilizer and watch it take off. If you made a fertilizer solution for your rose bush by pulverizing dried crickets, mixing them in water, and spraying the solution on your plants, would you have a carnivorous rose? In fact, root fertilizers for plants such as palms do often contain ground-up crickets, oyster, and crab shells. Do you then have an insectivorous or seafoodivorous or—God forbid—shellfishivorous palm tree? Maybe; maybe not.

The confusion over which plants are or are not carnivorous stems from how we define the term. It has been generally assumed that to be called carnivorous, a plant needed to do several things: lure prey, somehow catch it, kill it, and then digest it, usually through the production of enzymes and acids used to dissolve the victim into palatable form. It's this "digesting" that's the controversial part.

Some plants are included under the heading of "carnivorous" even though they don't produce actual digestive enzymes. *Darlingtonia* and *Heliamphora* are two examples of pitcher plants that apparently rely on bacterial action and other life forms to dissolve their prey. This "digestion by proxy" can make the determination of whether a plant is carnivorous rather complicated. If we simply defined a carnivorous plant as one that possibly benefited from the absorption of minerals obtained from captured and killed animals, instead of several hundred species, we might be talking about many thousands!

Petunias catch and kill insects. So do potato plants, tobacco, rhododendrons, and teasels. But they do so, it is believed, for defensive purposes. Plants like petunias are covered with sticky hairs, which make life difficult for insects like aphids who wish to eat them. Many insects become caught in these hairs and die. Potato plants also are covered with hairs. If an aphid breaks one of these hairs, a glue is secreted that cements the aphid to the spot. The common teasel has leaves that form "cups" where they join the stem. Rainwater collects in the cups; insects fall in, drown, and eventually dissolve. All of these plants, at one time or another, were viewed with suspicion by scientists studying carnivores. All of these plants probably absorb some nitrogen or other trace minerals through their leaves as the insects decompose. The rest of the minerals are possibly taken up through their root systems after rain. But they are not considered carnivorous because they lack the process of digestion.

Perhaps our definitions should be revamped. Many plants might fall into the category of "subcarnivorous" or "paracarnivorous." Take, for example, the long-standing problem of *Roridula.* The two species of this genus grow in South Africa. They look and behave so much like sundews that they were originally included under the genus *Drosera.* Their leaves are covered with sticky glands that catch enormous amounts of insects. But earlier in this century *Roridula* was excluded from the carnivores because it did not produce digestive enzymes. And so it stood until the 1990s, when two surprising discoveries were made. The first was by Steve Williams. Through DNA research, he discovered that *Roridula* was more closely related to *Sarracenia,* the pitcher plants on the other side of the Atlantic, than they were to *Drosera.* He jokingly suggested that *Roridula* be included in the family Sarraceniaceae and that its carnivorous nature be reexplored. It was, which led to the second surprise. It has long been known that *Roridula* plays host to a curious insect called the assassin bug. These bugs live on the plant, and for reasons still unknown can traverse the sticky glands with no problem at all. When other insects become caught, the assassin bugs close in, stab their needlelike mouths into the struggling prey, and suck out the juices. Alan Ellis and Jeremy Midgely discovered the amazing reason for this cooperative venture. Assassin bugs, after sucking dry *Roridula*'s prey, secrete

Roridula gorgonius covered with prey—but is it carnivorous?

a nutritious substance onto the plant that the leaves then absorb—a true carnivore by proxy! *Roridula* plays host to assassin bugs that act as a "surrogate gut." This discovery led Oxford carnivorous plant specialist Barry Juniper to comment on petunias, potatoes, and tobacco: "They're all killing machines. I wouldn't be surprised if they absorb decayed products from their prey."

How have such plants evolved? Definite theories on the evolution of carnivorous plants are few; the almost complete lack of fossil evidence coupled with the current shifting of ideology among evolutionists may make theorizing an exercise in futility. Uniformitarianism, or gradualism, as popularized by Darwin, Wallace, and other nineteenth-century scientists, holds that evolutionary change in both biology and geology is a very slow progression of events that occurs even as we speak. Darwin's theory of the origin of species by natural selection relied on rare and random mutation giving rise to new traits that, if beneficial to the species, allowed it to compete better among its peers and pass those traits down to its offspring. Darwin's dream that the fossil evidence for such transitional forms was simply missing from the evolutionary record has turned out to be mere wish fulfillment, for most scientists today agree that there are few transitional forms.

Thus, under gradualism, Darwinists imagined how a basic oval leaf slowly evolved, step by step, into the simple, rolled up, funnel shape of something like a *Heliamphora* leaf; then through random, accidental mutations over eons of time, the plant added genes to eventually produce the drug coniine in *Sarracenia flava* or the light windows of *Darlingtonia* or the symbiotic relationship of *Roridula* and assassin bugs . . . well, okay, maybe it's not so easy to imagine! Or, in the words of Francis Lloyd, who in his 1942 book *The Carnivorous Plants* mused on how the complex trap of *Utricularia* could have possibly evolved under gradualism: "Since we cannot answer these questions, it is perhaps as well to say no more."

A butterwort snacking on fungus gnats.

Currently, beliefs in gradualism are eroding. Scientists are realizing that for long periods of time species of life on earth are stabilized, with little or no evolutionary progress. Then, periodically and very suddenly, geological and biological changes take place. Older species suddenly vanish, while new ones appear quickly with no recorded transitional forms. Others remain unchanged. While research in areas such as DNA may lead to conclusions concerning relationships between species, including carnivorous plants, how those species actually evolved is still the deepest of mysteries. I will mention some evolutionary theories in the genus chapters to follow. The answers may come over the next century—maybe through the theories of punctuated equilibrium whereby mysterious global catastrophies killed many species and sped the evolution of new ones, as proposed by the late Stephen Jay Gould in his opus *The Structure of Evolutionary Theory*. Or perhaps Derek Ager was correct in his *The Nature of the Stratigraphical Record*, in which he proposed that global cataclysms created the fossil records we know, leaving vast intervals of evolutionary time largely unrecorded. Many other intriguing theories concerning how a species can mutate into a new species abound, from "epigenetics" to "morphic resonance" to the "electric universe" ideas. Or perhaps a clearer understanding of the mystery will come from new ideas that no one has thought about yet.

The invention of the greenhouse in the early 1800s and its growing popularity during the Victorian era among Europe's upper classes allowed, for the first time, exotic plants from around the world to be successfully grown under controlled conditions. Commercial nurseries were developed to cater to the demand for exotic plants. Some of these firms, such as the famous Veitch and Sons (later Veitch Nurseries) in England, financed expeditions to far-off lands around the world to collect unusual plant life. Their fanciful catalogs in the 1800s boasted palms, orchids, hoyas, succulents, and carnivores. Among the most popular insect-eating plants offered were the *Nepenthes*, but such firms also sold *Sarracenia*, *Drosera*, *Dionaea*, and *Cephalotus*. Breeding programs developed showy hybrids that competed—along with roses and orchids—for prestigious awards at the Chelsea Flower Show. Magazines like the *Gardeners' Chronicle* popularized the plants and offered cultural techniques. Exotic plants were usually very expensive and during the 1800s were affordable only for the very rich. But hardly a greenhouse existed on the estates of the wealthy that didn't have *Nepenthes* hanging from the rafters along with the palms and the orchids.

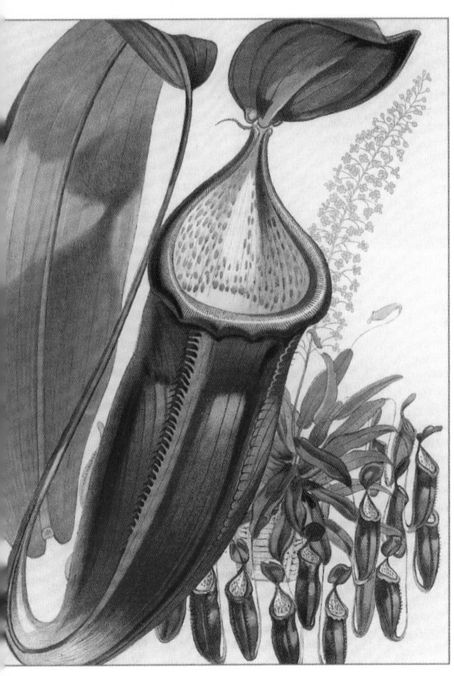

Tropical pitcher plants, such as this *Nepenthes sanguinea*, were popular greenhouse specimens in Victorian England.

World War I changed all of that. The shortage of fuels to heat the greenhouses caused the sudden death of vast botanical collections. Many prized cultivars and species disappeared forever. Only here and there, hidden away in public botanical gardens and universities, did some of the plants survive.

It wasn't until after World War II that the hobby of growing carnivorous plants began to make a modest comeback. During the war Francis Lloyd published his scientific work *The Carnivorous Plants* in America, the first such work since Darwin's 1875 *Insectivorous Plants*. In Japan, where the love of cultivating exotic plants (such as dwarf rhapis palms and bonsai) suffered heavily during that war, the first carnivorous plant society was begun in 1948. It still thrives.

During the fifties and sixties, the cultivation of ornamental plants steadily increased with the booming economies worldwide. But except for the efforts of a handful of interested individuals, carnivorous plants remained obscure to the general public. Venus flytraps, dug up out of their native North Carolina habitat and sold as an occasional novelty, were the only insect-eating plant available on the mass market. No popular literature on the plants appeared except for infrequent articles in magazines such as *National Geographic.*

The modern carnivorous plant hobby began in earnest during the seventies. Two hobbyists, Joe Mazrimas in California and Don Schnell in North Carolina, began to communicate with other collectors around the world. This developed into the International Carnivorous Plant Society (ICPS) and its publication: *The Carnivorous Plant Newsletter.* For the first time, enthusiasts had an organized format to exchange cultural information as well as seed and plant material from around the world.

The seventies and eighties also saw the publication of several books on the subject, as well as a few nurseries and collectors that began to take the ecology of carnivorous plants seriously. Instead of removing plants from their rapidly diminishing native habitat, they began to propagate them. Three nurserymen who helped distribute hundreds of rare, sought-after carnivores were Adrian Slack in England, Marcel Lecoufle in France, and Bob Hanrahan in the United States. Plants that many hobbyists had never dreamed they would see in their lifetimes were suddenly becoming available.

The nineties saw a steady increase in the popularity of carnivorous plants. Smaller local societies, usually associated with the ICPS, have sprung up in various parts of the United States, Europe, and Australia. Tissue culture propagation has made many once-rare plants common and affordable. Numerous nature programs on cable television are making carnivorous plants more visible to the general public. The plants are beginning to appear more often at flower shows and public botanical gardens. General nurseries are retailing a wider variety than just the familiar Venus flytrap. In 1998, the first edition of *The Savage Garden* was published, receiving rave reviews and winning two book awards, including a book-of-the-year award from the American Horticulture Society, and subsequently has sold tens of thousands of copies. This was followed by the explosion of websites, forums, and commercial nurseries on the Internet, which has spread interest in carnivorous plants to every continent of the world.

So how does one use this book?

I have written and revised *The Savage Garden* as a practical guide to growing carnivorous plants (which, for the sake of brevity, I will refer to as "CPs" throughout the rest of the book). It is divided into three main parts: Part One covers the primary points of cultivation, including hard goods. Part Two outlines the various places you can grow CPs, from greenhouses to bog gardens. Part Three introduces all the popular genera of insect-eating plants, as well as their history, habitats, and habits, including their specific cultivation requirements. This is followed by a brief section on resources.

PART ONE

THE BASICS OF CULTIVATION

Sarracenia flava var. *rubricorpora* with *Drosera filiformis* var. *tracyi* and *Pinguicula planifolia* in an open, wet, sunny habitat in the Florida Panhandle. Some introduced Venus flytraps are in the background.

Dr. Larry Mellichamp of the University of North Carolina once told me, "Whenever I have seen carnivorous plants in the wild, whether it was in South Africa, the Florida Panhandle, or Northern California, the habitats were often strikingly similar." Broadly speaking, this is a rather true statement, although of course some CP habitats are vastly different from others. I have been struck by the similarity of pygmy forest bogs in Mendocino County, California, to those I had explored as a youngster clear across the continent in the pine barrens of New Jersey. If someone were to show me photos of each, they would be virtually indistinguishable. In fact, in a small boggy area of the pygmy forests near Fort Bragg on the Northern California coast, many carnivorous plants from around the world have been introduced, much to the annoyance of ecological purists. Although the only native carnivorous species is *Drosera rotundifolia*, plants that have been introduced and have been growing well for many years include Venus flytraps from North Carolina; many *Drosera* species from South Africa, Australia, and Asia; *Sarracenia* from the southeast of North America; *Pinguicula* from Europe; and *Darlingtonia*, whose native habitat is found a few hundred miles to the north of Fort Bragg. Even *Heliamphora*, from Venezuela, survived—until the plants were stolen!

This example illustrates that many CPs grow in habitats so similar to one another that cultivating the plants is simplified by an understanding of a few general facts. Carnivorous plants typically inhabit wet, low-nutrient soils through which water may be slowly moving. This moving water usually carries away what minerals there are in the soil, which explains why it tends to be low in nutrients. The soil is often sandy, with a ground cover of patchy mosses such as sphagnum, which turns into peat as it ages. The habitat is often sunny, and the few trees are commonly stunted by growing in the infertile soil. Pines or other evergreens, whose needle droppings may further add to the soil's acidity, are the most common trees in such habitats. Grasses are also common. In short, a wet, low-nutrient, sunny environment is preferred by most CPs. The only major difference between habitats, usually, is the climate.

Soils

Specific soil recipes will be offered in the section on genus cultivation. Here I will discuss the individual ingredients used in most artificial soil mediums for carnivorous plants.

Sphagnum Peat Moss

Peat moss is probably the most important soil ingredient for most CPs. It must always be sphagnum peat moss; the word "sphagnum" must always appear on the moss packaging. Peat moss is usually sold in packages ranging from small bags to large, dried, bricklike bales. It is commonly of Canadian, Irish, or German origin. It is a fibrous moss, with a consistency close to that of a fine sawdust, and from light to dark brown. It is available in most general nurseries, usually as an additive for garden soils. It is what many carnivorous plants grow in naturally. Sphagnum peat is very acidic, usually with a pH between 3 and 5. (On a scale of 1 to 14, a pH of 7 is neutral.) It can hold as much as ten times its own weight in water.

Peat moss should be broken up until it is similar to sawdust before it's used, with all lumps and clods gone. It should then be mixed with water until it resembles a soft, wet mud. Then it is ready for use. Avoid sedge peat or Michigan peat, which are entirely different substances. If the package doesn't say "sphagnum," don't use it! One problem with the commercial sale of sphagnum peat moss is the addition of fertilizers such as the Miracle-Gro brand to the moss, because CPs need low-nutrient conditions to thrive. Read the labels carefully and avoid any peat to which fertilizers have been added.

Long-Fibered Sphagnum Moss

This is a problematic soil ingredient for two reasons. The first is that what is sold in most nurseries is often not long-fibered sphagnum moss at all, but decorative green moss or "Oregon sheet moss." "Sphagnum moss" has unfortunately become the generic term used for any dried, fibrous moss used in horticulture. Decorative moss is often used as a basket liner or topdressing for potted plants, as is true sphagnum. Worse, I have seen packages of decorative green moss actually being sold as

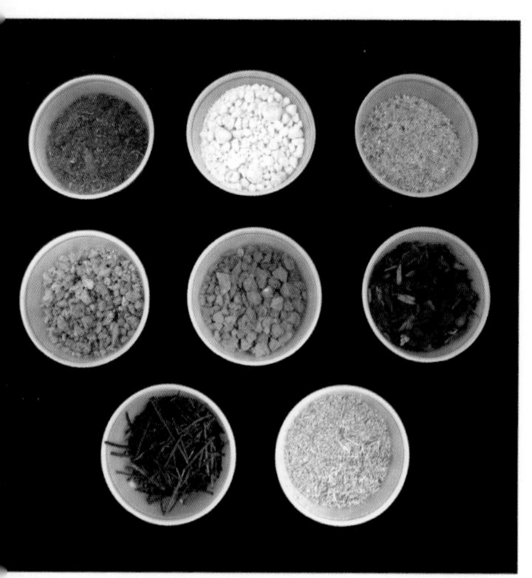

Soil ingredients for carnivorous plants. Left to right, top: sphagnum peat moss, perlite, sand. Middle: pumice, lava rock, orchid bark. Bottom: tree fern fiber, milled sphagnum moss.

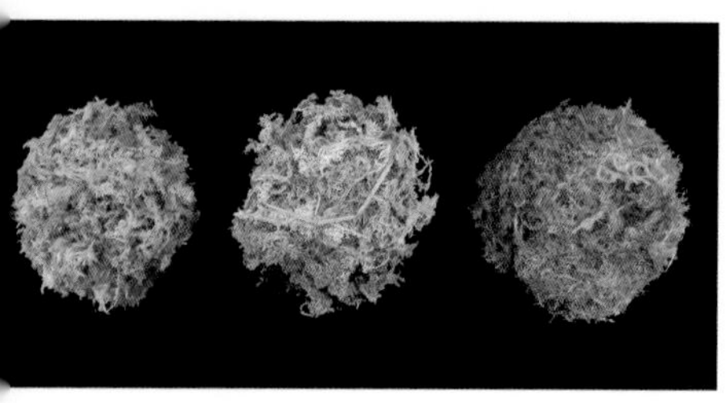

Left to right: long-fibered New Zealand sphagnum moss, domestic American sphagnum moss, and green decorative moss (lethal to carnivorous plants as a soil).

sphagnum moss. I have seen many CPs killed because decorative green moss was mistaken for true sphagnum.

Further complicating the situation is the confusion between long-fibered sphagnum and sphagnum peat. Sphagnum moss, usually greenish or reddish, grows along the surface of the moss bed in a typical bog. It grows in long, ropey strands, with the growing head of the strand at the bed's surface. These strands can be rather lengthy, extending deep below the surface of the growing tips. Only the first 6 to 7 inches (15–18 cm) at the surface are colorful and alive. Underground, the moss turns brown, and a couple of feet (0.6 m) below it decomposes into sphagnum peat moss. The peat may extend quite a way underground and may be hundreds of years old. Testifying to the sterile conditions of the moss is the fact that centuries-old, virtually unspoiled human bodies have been found deep within peat bogs in Europe.

Long-fibered sphagnum usually refers to the dried, ropey strands collected from the moss bed's surface. Sphagnum peat is the decomposed moss harvested from deep underground. In soil recipes for most carnivorous plants, the more reliably labeled peat moss is the preferred ingredient.

For some carnivorous plants that prefer long-fibered sphagnum moss in their soil recipies, such as the *Nepenthes*, high-quality moss from New Zealand or Chile is the best and available most often through CP or orchid specialists. In the United States, what is known as "domestic" sphagnum, harvested in states like Wisconsin, is a less expensive type that often includes twigs and leaves. While inferior as a soil, it is excellent to cover the holes of drained pots to keep peat soils from slowly leaking out.

Live Sphagnum Moss

Living sphagnum is a beautiful moss—lush, billowy, and colorful when well grown. Many carnivorous plant growers love the sight of bright green or red sphagnum covering the soil surface of a potted CP, even if it isn't used throughout the whole pot. Live sphagnum can be induced to grow as a topdressing to peaty, wet soils, although sometimes with difficulty.

Live sphagnum moss.

The drawbacks to live sphagnum are many. First, it is difficult to obtain commercially. Second, it grows slowly. Third, when it does grow, it will easily cover up small plants like Venus flytraps, causing them to rot. Fourth, it is difficult to maintain. Hot sun will burn the growing tips, and the lightest application of fertilizers or minerals can cause algae growth and death to the moss—usually. A wonderful development in recent years at California Carnivores was the discovery of the Maxsea brand of fertilizer. Unlike with orchid or epiphytic fertilizers, we have found that Maxsea not only doesn't kill live sphagnum moss but also accelerates its growth phenomenally. Read more about Maxsea in the Fertilizing Carnivores section ahead.

Among my own plants I have found that the only reliable place live sphagnum will grow for the long term is as a topdressing in pots of highland *Nepenthes* or similar genera or in bog gardens that receive frequent overhead sprinkling with purified water. Live sphagnum will not survive for more than a year in hothouses, because the moss usually grows in cool alpine climates or climates with cold winters.

Live sphagnum will also usually grow well outdoors in climates that are humid and receive lots of rain. It will succeed in very wet bog gardens and simulated similar container environments; for instance, undrained tubs of tall *Sarracenia*. Since this is a live "soil" that grows, trimming will be needed to keep it from crowding smaller plants.

Milled Sphagnum

This is dried, long-fibered sphagnum moss that is shredded to the consistency of fluffy, coarse sawdust. Usually sold as a medium on which to germinate seed, and resistant to the dreaded damping-off fungus that attacks seedlings in damp environments, it would be excellent as a peat substitute were it not so expensive.

Sand

All sand used in soil recipes for carnivorous plants should be sand that has been well washed. Nurseries and garden centers usually have washed sand available for use with potted plants. Washed "play sand" meant for use in children's sandboxes is also good, and is often sold in plant nurseries and home supply stores. Avoid self-collected beach sands or river sands, which are often contaminated with mineral salts.

Perlite

Perlite is a mineral rock that is heated until it expands, creating a lightweight, granulated soil additive that will hold both water and air. Usually white, it is available at garden centers. Fine and medium grades are preferred for most CPs. Commercial nurseries generally prefer to aerate the soil in potted plants with perlite rather than sand, because sand is very heavy and leads to high shipping costs. However, most perlite is also slightly alkaline (pH of around 8), so never use more than 20 percent in the soil mix for acid-loving CPs.

Pumice

Another lightweight, airy rock used in horticulture, pumice is usually gray.

Lava Rock

This volcanic rock is reddish brown, and available at most garden and landscape centers.

Vermiculite

This is a mica, processed similarly to perlite and serving a similar purpose. It is usually golden-brown. Since most vermiculites contain some minerals (such as magnesium and potassium), and it breaks down, turning to mush, I no longer use it for CPs.

Orchid Bark

Available at most garden centers, this is usually the bark of evergreen trees. Popular for orchids, it is helpful in soils for *Nepenthes*, among other CPs, and makes a decorative topdressing on peaty soils used for plants such as *Sarracenia*. A fine-grade bark is preferred by CPs.

Shredded Bark

Similar to orchid bark, this is shredded to the consistency of confetti.

Osmunda or Tree Fern Fiber

This natural material is usually blackish brown and has a consistency not unlike that of broken toothpicks. It is used to aerate soils.

Rock Wool

This interesting substance is sometimes used for plant propagation, especially in cases where stem cuttings are used. Rock is liquefied and spun into lightweight fibers, then usually molded into small bricks. Difficult to find in the general nursery, it can sometimes be obtained through specialty houses. Bricks of rock wool retain moisture, and stem cuttings of *Nepenthes* root well in it. The entire brick is later planted in soil.

Coconut Soils

In recent years coco peat, or coir, has appeared on the market, produced as a by-product of the coconut palm industry. I had disastrous results with this product when it first appeared on the market—primarily due to salt contamination. While some coco peat or coco chips are now claimed to be heavily washed in fresh water and some growers use it, I don't.

Water

This is probably the single most important issue concerning the cultivation of carnivorous plants. Generally speaking, CPs require water that is low in dissolved mineral salts, and they usually need lots of it. Dissolved minerals are usually indicated on analysis as p.p.m., or parts per million, or t.d.s., meaning total dissolved solids. Water used for the cultivation of carnivorous plants is safest for the plants when it is below one hundred p.p.m. of dissolved solids, and the lower the better. It is the use of hard, mineral-laden water that most often causes carnivorous plants to decline in cultivation. Since the majority of CPs are grown in basically undrained conditions, constant use of hard water continually adds minerals to the soil. Since CPs are adapted to grow in low-mineral soils, roots will begin to rot and the overall health of the plant will suffer when dissolved minerals are left behind as water evaporates or is absorbed by the plant. The more you water, the more minerals you add to the soil.

Water with a t.d.s. count of 160 p.p.m. can be suitable for most CPs if they are outdoor plants that get frequent or seasonally heavy rainfall, which helps leach mineral buildup from soils and water trays, or are grown in greenhouses not sitting in trays but watered from overhead—such as the *Nepenthes*. Rinsing water trays periodically also helps eliminate mineral buildup.

Small battery-powered t.d.s. meters to test water for dissolved solids are commonly found for sale on the Internet (try www.ebay.com) or at aquarium and hydroponic stores.

Most drinking water is fairly neutral in pH (7.0); however, limestone rock can cause groundwater sources or wells to be too alkaline, such as in England or Florida. Occasionally, water companies will add lime to acidic water to slow the corrosion of pipes. San Francisco is one example of this. If you are unsure of your water's pH, you can get simple test strips at aquarium and hydroponic supply stores or online. Water is too alkaline if its pH is above 8.0.

Some public water systems supply water that is naturally low in minerals. Your public water quality depends, of course, on where you live and where your water comes from. Your water company can supply you with an analysis of your tap water. Some folks mistakenly believe that boiling water or allowing it to sit out in an unsealed container for a day or two will allow the minerals to magically disappear. This is not true. The only thing that will vanish from the water with such treatments is the dissolved chlorine gas that is added to water to kill bacteria. The minerals, unfortunately, are left behind.

If you have hard tap water, the best water to use for carnivorous plants is collected rainwater or water that has been demineralized or purified. If you live in an area of frequent rainfall on a year-round basis, you can collect this water for use with CPs. Make sure, however, if you collect rainwater from the gutters on the roof of your house, that your roof has not recently been treated with fire retardant, moss inhibitor, or other chemicals. In other situations, it is best to buy purified water or purify it yourself. For a small selection of plants, such as those in a terrarium under grow lights, water demand may not be high, so providing purified water from a grocery store or water machine may be affordable.

Commercially bottled water has become a hugely profitable business, as its producers have promoted the idea that bottled water is safer than tap water; however, municipal water is actually subject to more stringent safety requirements. Furthermore, bottled water that is not specifically identified as spring or well water is likely to be tap water that may have been purified and had minerals and salt added to improve the taste (purified or distilled water can taste rather flat). So if you buy bottled water, make sure the water is distilled or the label specifically states that it is low in sodium. Do not use mineral water, mountain water, or spring water, unless the label states it has been purified and has zero sodium content.

If your water use for CP cultivation exceeds more than a couple of gallons (7 liters) a week, the only answer is to purify your own. Distilleries are impractical for home use—a lot of fuel is needed to boil water to evaporation and recondense it as distilled water. Much more practical is a reverse-osmosis unit to produce your own purified water. Reverse-osmosis (R.O.) systems used to be expensive and hard to find, but in recent years prices have gone down dramatically, and under-the-sink units are available in most large home-supply and hardware stores and at some specialty plant nurseries.

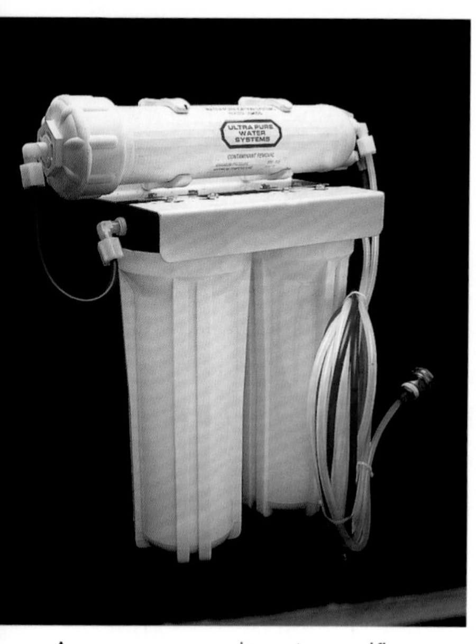

A reverse-osmosis water purifier.

Most under-the-sink units run tap water through particulate canister filters, then through the R.O. membrane, which "squeezes" the water through a fine micron filter, separating the good, purified water from the bad, mineralized water. The former is collected in a tank under the sink, while the latter goes down the drain. The collected purified water, about 99 percent free of dissolved solids, is usually accessible through a separate faucet on the sink. A typical system of this sort will produce between 3 and 5 gallons (11–18 liters) a day of pure water.

No-frills R.O. systems can be purchased online; these cater to people who need pure water for plants, such as orchid and CP growers. These units are hooked up to hose bibs in the garage or a protected place outdoors. With these simplified units, users collect the purified water in their own bottles, water tanks, or clean plastic waste cans. The wastewater is either allowed to go down a sink drain or run out via tubing for use in the general garden. (Most garden plants are unaffected by the concentrated minerals, and rain will leach them away.)

The Water Tray Method

Since most carnivorous plants need to grow in soils that are permanently wet, the easiest way to accommodate them is to grow the plants in pots that have drainage holes at the bottom and set the pot in a saucer or tray that always has some water in it. The majority of CPs are happy when approximately 1 inch (2.5 cm) or so of water is maintained in the saucer at all times.

There is a wide variety of containers that can be used as a water tray, and unless you are around your plants all the time, or have a dependable person to water your plants when you are away, I recommend you search for rather large saucers to accommodate your plants.

In the greenhouse, most collectors use large, undrained seed flats that can hold several pots and are around 2 inches (5 cm) in depth. For single pots, a regular saucer may be used, but it should be one or two sizes larger than you would usually provide for the pot size, so it will hold as much water as possible. There are plastic saucers available that are extra deep, often as deep as the pot; these are excellent for most CPs. Extra-deep saucers are often used for houseplants whose pots sit in decorative baskets. They make great containers for pots of CPs because water can be flooded to the top of the pot and allowed to gradually evaporate before you add more. Not only will you have to add water less frequently, but also this technique mimics the habitats of those carnivorous plants that live in soil waterlogged after heavy rains. As the water table drops, oxygen will permeate the still-wet soil.

So-called self-watering pots, sometimes used for plants such as African violets, are generally unsuitable for CPs. These pots work with wicks set into a water tray. They will keep the soil lightly dampish, which is fine for most houseplants but not wet enough for most CPs.

Avoid using clay saucers, unless they are glazed and therefore waterproof. Plastic water troughs can be decorative and large enough to hide the pots themselves. Busboy and dish tubs may not be attractive, but they can serve the same purpose. The same goes for clean kitty litter trays.

Pottery and Containers

Generally speaking, plastic pots are usually best for carnivorous plants, but as we will see in the later section on genus cultivation, there are exceptions to this rule. Terra-cotta clay pots are usually avoided for several reasons. Evaporation will be high through the porous material, and clay can absorb harmful mineral salts that may accumulate in the soil. Also, since the soil is usually kept rather wet, unsightly and slippery algae and mosses may grow on the pot's surface. However, you will see in Part Three, on individual varieties of CPs, that there are a few plants that actually prefer terra-cotta clay pots.

On the other hand, clay pottery that has been glazed makes very attractive and suitable containers in which to grow CPs. Glazed pots may be of the type with holes in the bottom, which can then be set into a saucer

Left: Containers for carnivores. Clockwise from upper left: *Sarracenia minor* in a plastic pot and saucer; *Darlingtonia* in a glazed ceramic pot and saucer; *Nepenthes villosa* in a wooden box; undrained glazed ceramic with *Drosera intermedia, D. filiformis* var. *filiformis,* and *Utricularia gibba*; *Pinguicula moranensis* in an undrained ceramic tea cup. Right: *Sarracenia minor* 'Okee Giant' in an undrained glazed ceramic bowl.

of water. But some CPs, such as *Sarracenia* and most *Drosera*, *Utricularia*, and *Pinguicula*, will also thrive in undrained pottery that has been glazed.

Although the tray system is the most common method used to grow carnivorous plants, it certainly is not necessary if you grow the plants in undrained containers. These containers should be waterproof and made of materials such as plastic, glazed clay, or glass. Containers that are wooden may be lined with a durable sheet plastic, thus making them both undrained and waterproof. When CPs are grown in an undrained container, the soil should remain wet at all times. The water table should be allowed to fluctuate somewhat, from waterlogged to damp, so that air can enter the soil.

Undrained containers are certainly not suitable for all carnivorous plants. *Nepenthes*, for instance, always require wet but well-drained soil, and Venus flytraps and the West Australian pitcher plant generally despise undrained containers unless they are particularly deep. In short, while almost all CPs grow in wet soils, the degree of that wetness may vary from genus to genus.

Natural Light

Only rarely will you find carnivorous plants growing in the wild in deep, dense shade. Although a few tropical pitcher plants may grow in the shade of a forest canopy, and several butterworts may be protected from sun by growing in shaded grottos of dripping, wet rock, the vast majority of CPs are found in open, sunny habitats. The occasional pine tree or tall grasses may offer partial shading in some instances, but for the most part sunlight is as important to most carnivorous plants as wet, acidic soil.

Let me be clear about what I mean when I use the terms "sunny," "partly sunny," and "bright shade" throughout this book. By "sunny," I mean that the plant will do well outdoors or in a greenhouse in full sun for most of the day, even when a shadecloth is applied. On a windowsill, sun should shine directly onto the plants for at least five or six hours. "Partly sunny" conditions means that direct sun should bathe the plant for at least two to four hours a day; the rest of the time it should be in bright shade. "Bright shade" means intense light, but direct sun should not hit the plant for more than an hour or so each day.

Photoperiod (the length of time a plant is in the light) is an important part of natural lighting. This light does not necessarily mean direct sun. Along the equator, the daylight/darkness period is more or less evenly divided, with roughly twelve hours of daylight and twelve hours of darkness. In a temperate zone, the daylight/darkness ratio changes from season to season as the earth, tilted on its axis, orbits the sun, so a *Nepenthes* from the tropics may receive an evenly divided day and night period through much of the year, whereas a Venus flytrap from North America may receive fifteen hours of light at the summer solstice and only nine hours of light at the winter solstice. This is very important when a grower is considering the dormancy or rest period of temperate carnivorous plants. Some people may think that cold weather is what makes a temperate species go dormant, but that is only half of the story. The length of the daylight period is the story's other half. Photoperiod also influences other functions in the life cycle of plants, such as flowering times and the formation of reproductive brood bodies, as in the pygmy sundews.

Dormancy

Dormancy in plants is rather similar to hibernation in animals. Many carnivorous plants go dormant during adverse seasonal conditions. Most temperate plants that grow in climates with cold winters and short daylight periods lose their leaves and stop actively growing during this time. As the weather warms up in spring and the photoperiod gets longer, the plant resumes its growth.

Dormancy can also take place for other reasons. As we will see with plants such as tuberous sundews from Western Australia, almost the opposite occurs. Some plants actively grow during the cool, wet winters and then go dormant as the days get longer, heat increases, and the winter-wet soils dry out.

Dormancy in carnivorous plants that require it must be respected and permitted to occur. Otherwise, the plant may die. A temperate Venus flytrap, for example, grown in simulated tropical conditions such as a warm terrarium with grow lights operating permanently on a sixteen-hour photoperiod will eventually get sickly and die.

Winter dormancy in temperate carnivorous plants generally requires two things: a shortening of the photoperiod and a cooling of the temperature. For plants requiring a summer dormancy, usually a drying out

Dormancy is crucial to many carnivorous plants. Left: American pitcher plants (*Sarracenia*) in winter. Top right: Dormant resting buds of temperate sundews (*Drosera*). Lower right: A Venus flytrap (*Dionaea*) in winter.

of the soil is crucial. Other varieties, such as most Mexican butterworts, change from a carnivorous habit during wet, hot summers to a noncarnivorous stage of succulentlike growth during the cooler and drier winter months, when most of the subtropical zone experiences a drought.

For the most part, if left alone and if grown in the proper environment, carnivorous plants will go dormant on their own. Allow them to do so; you should never force a carnivorous plant into growth during a season when it should be resting.

People who live in tropical places often wonder if they can grow temperate-zone plants there. The answer is yes, *sometimes*, if some extra effort is made to accommodate a plant's cool rest period. An American pitcher plant, for instance, grown outdoors in a warm, subtropical place like Hawaii, may need to be uprooted, put into a plastic bag, and refrigerated during the winter. If the Hawaiian gardener lives in the cooler highlands, the temperature drop and shorter daylight period in winter may be sufficient to carry his or her plant through its winter rest period. Cool basement windowsills may offer a similar environment.

Artificial Light

You live in a fifth-floor apartment in, let's say, Quebec, Canada, and all of your windows face north and are sunless. You have no patio to summer-grow your plants, and the fire escape is in permanent shade. Can you still grow carnivorous plants? Yes. In fact, even if you live in a dark basement or a densely shaded house in a redwood forest, carnivorous plants can thrive even in the darkest of corners. The solution is to grow them under artificial light, which can be done with or without a terrarium, although most growers use tanks or terrariums to grow their plants indoors. If the humidity in your home remains above 50 percent, many carnivores will do fine under grow lights out in the open and not need a tank or aquarium. However, most CPs appreciate the higher humidity provided by such enclosures, especially the tropicals. We will discuss this further under the section on terrariums, beginning on page 60.

In recent years grow light technology has greatly advanced thanks to hydroponics and the marijuana industry. Once-popular metal halide and high-pressure sodium lights use a lot of electrical power and give off a lot of heat. Thus a return to much-improved fluorescent lighting is what I generally recommend for carnivorous plants.

The best grow lights to use are called T-5 High Output fluorescents. They contain two to six bulbs and conveniently come with hoods measuring 24 inches (60 cm) and 48 inches (120 cm) in length. These fit atop both of the most popular aquarium sizes growers use, which are 20- or 55-gallon (roughly 75–208 liters) aquariums. Usually the former tanks are 24 inches (60 cm) long and the latter 48 inches (120 cm) long. You can find these lights for sale at hydroponic stores or even pet shops or do an online search for "grow lights."

There is a common misconception about how far below the fluorescent lights the plants need to be. The lights should generally be no farther than 12 inches (30.5 cm) above the pots. Naturally, this means that tall plants like *Sarracenia* are usually unsuitable for fluorescent lights (although low-growing species are an exception) and that when *Nepenthes* begin to grow climbing vines they will have to be pruned back or moved to a greenhouse, unless they are the smaller growing species.

Some other grow lights (which I have not personally used) are quite suitable for CPs. Dual-spectrum compact fluorescents measure only 16 inches (40.5 cm) long for smaller areas. Some hobbyists have had excellent luck with the new LED (light-emitting diode) lights. There are even large "grow-tents" with high-output lights that can be used in areas like a basement—a greenhouse underground!

Feeding Your Plants

"I gotta find food for Master. Food I gotta find for Master.
For Master I gotta find food."

—SEYMOUR KRELBOYNE, *Little Shop of Horrors*, 1960

In horticulture, feeding one's plants usually means the application of fertilizers. With carnivorous plants, of course, we are speaking literally!

If you are growing your plants outdoors, feeding them certainly won't be necessary. CPs will lure, catch, and eat numerous insects outdoors on their own. Flies, ants, gnats, moths, beetles—any prey similar to those they lure in their natural habitats will be attracted to and fed on by the different plant species. Flying insects usually are the most common victims. But even ants won't necessarily be saved by the water trays that surround the pots. Older trumpet leaves may lean over and touch the ground, providing a bridge for ants to cross. Some folks I know purposely put twigs or sticks across water barriers to entice crawling insects to visit their potted plants—a visit the insect may later regret.

Be warned, however, that bridges to pots set in water trays will also allow certain pests access to your plants, and a slug may eventually be caught in a flytrap only after having munched a few holes in a tender and newly emerging *Sarracenia* leaf.

Outdoors, some carnivores will catch such large quantities of insects that the result can be startling. A sundew's leaves may be black with gnats; every flytrap leaf may be shut tight on flies; and American pitcher plant trumpets may topple from the weight of hundreds, if not thousands, of prey. The pitiful buzzing of trapped yellow jackets or flies can be unsettling to some people. The sensitive should never peer down the tube of a trumpet plant that is choked with ants and flies. The ants, made quite insane by their predicament, will be merciless toward the helpless housefly that tumbles into their madhouse prison. At times I have been so disturbed by their suffering that I have freed ladybugs and even yellow jackets from an agonizing death. Their plight appears not unlike that of a human being trapped in a greased well filled with a horde of rats.

An autopsy on *Sarracenia* proves they are gluttonous pigs.

And speaking of ants, I have a warning: ants in a greenhouse will not only provide ample food for American or tropical pitcher plants, but they will also sometimes begin to cultivate some of their own food, namely scale insects and aphids, from which the ants extract honeydew. This may be a revenge tactic of the ants: plant eats ants, ants farm

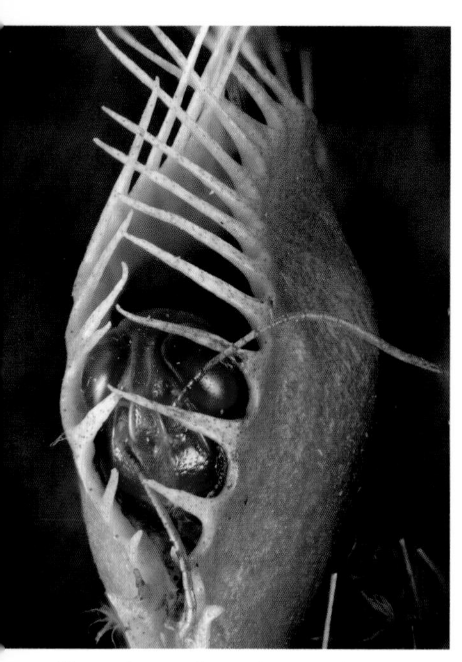

The unhappy face of a grasshopper as it is digested alive.

scale on plant, scale sucks juices from plant, ants feed on scale honeydew. Nature works in funny ways.

Ants in the greenhouse are often of the nomadic sort. A queen may set up a nest in a potted *Nepenthes* and then wonder where all her workers are disappearing to. (I have never found ant nests to do any particular damage to the roots of a potted CP.) When the pot is watered, the whole nest swarms in a panic, carrying eggs and pupae out of the deluge. They either move the nest to another pot or return when the water drains.

If you have ant nests in your potted greenhouse plants, keep an eye on them. At the first sign of scale, you will have to use an insecticide, or you may want to discourage the ants by laying a few flea collars around the infested pots to keep the ants away. You'll find more about this under the section on pest control, beginning on page 38.

For plants grown in insect-free areas, you will have several possibilities for how to go about feeding them. One method is to simply hand-feed them. Forceps or tweezers are helpful: if you catch a fly yourself, it is usually much easier to introduce the doomed insect into the maw of a Venus flytrap by means of forceps, because you have to stimulate the trigger hairs within the trap; you cannot simply drop the fly in. Often the fly escapes just as the trap closes, which is as cruel to the plant as stealing candy from a baby. Be sure to wash your hands after handling germ-ridden flies, and urge children to do likewise—or feed plants cleaner food such as sow or pill bugs.

Many CPs eat tiny insects. You can gather small ants from an ant trail on a sidewalk with a damp paper towel and drop them into a paper cup. Your neighbors may think you're strange, but if they know you grow carnivorous plants, they probably think you're strange anyway! You will have to separate the ants from the grains of sand and other debris. Then you can sprinkle the ants on plants such as sundews or butterworts.

Alternatively, some windowsill plants can be placed outdoors temporarily to catch their own food. Be careful not to place them in an area hotter or sunnier than their normal environment, or the plant may burn or go into shock. Shade is best for a windowsill plant placed outdoors for "the hunt."

Aquatic bladderworts typically feed on minute swimming things such as daphnia (water fleas). You can collect these in almost any pond or lake with a paper cup, then add the contents to the water bowl where you grow your *Utricularia*.

Perhaps the easiest way to feed most carnivorous plants grown in an insect-free environment is to visit your local pet shop. Here you will usually find a great assortment of insect food, particularly if the shop caters to reptile and amphibian fanciers. Live crickets, from pinhead-sized newborns to adults, can be used to feed various plants, from Venus flytraps to pitcher plants. Wingless fruit flies can be fed to sundews, butterworts, rainbows, and other sticky plants. Mealworms will drown quickly when dropped into various pitcher plants, but make sure they don't escape, or they could infest your soil.

This staghorn sundew is a graveyard of insects.

You can also freeze these insects for future use, whether you purchase them or catch them yourself. If, despite your keen interest in raising carnivorous plants, you find that feeding your plants live insects makes you feel ill or guilty, there is also an assortment of dried insect food available at good pet shops, which most carnivorous plants readily accept. Some are even vitamin fortified! Dried flies, musca larvae, and ant eggs will save you much fuss and bother. Even plants like sundews, which normally require some moving stimuli to activate the feeding process, will soon curl around and drool over a dried insect applied to the leaf.

Yellow trumpets love hideous bugs.

Carnivorous plants will also sometimes eat human food. (I am not suggesting you feed your plants humans—that is highly illegal. . . .) Some CPs will accept and even drool over tiny bits of raw hamburger, cheese, powdered milk, and even chocolate (female plants). But these food products may be harmful to your plant, especially if they are overfed. A sundew leaf may curl around a bit of Hershey's chocolate, soon secreting digestive juices and making a pig of itself—but a few days later, mold or fungus may set in. So it is best to avoid such food as a regular part of the diet.

Do carnivorous plants really need to eat insects for their health and well-being? The answer, in my opinion, is more of a yes than a no. Insect food is highly beneficial to most CPs in that it provides them with the extra nutrients they need to flower, set seed, and grow larger each year. When deprived of such nutrients and minerals, the plant's health will gradually decline over time. It may not flower or set seed. The following year the plant may grow weaker, and the year after that weaker still.

Some species differ; a *Sarracenia*, for instance, when going dormant in autumn, sends all of those minerals from the thousands of insects it has caught in its leaves down into its rhizome for winter. The rhizome swells, more offshoots develop, and more flowers are sent up in spring. The stored energy has to go somewhere!

This does not mean you have to feed your plants on a daily basis during the growing season. A once-a-month feeding schedule from spring through autumn will usually do the trick. A few houseflies for your flytrap or a dozen or so crickets for your pitcher plants will usually provide the plants with the minerals they need for good growth. But you will probably notice that your healthiest plants are the ones that have eaten the most!

Fertilizing Carnivores

The basic rule of green thumb concerning the application of artificial fertilizers on carnivorous plants is this: strongly dilute the fertilizer and apply it as a foliar feed. It is well known in horticulture that most non-carnivorous plants can absorb minerals through their leaves; spraying the foliage of plants with a fertilizer can be as effective as feeding through the roots. This is, of course, also true with carnivorous plants. Since most of their leaves are especially adapted to absorbing minerals through specialized digestive glands, artificial fertilizers can be as readily absorbed as insect prey. By applying fertilizers on the leaves and avoiding drenching the soils, you won't have to worry about changing the low-mineral content of the plant's preferred nutrient-poor medium.

I will offer specific fertilization recipes in Part Three of this book; here I will only suggest some guidelines. First, if your plants are receiving a steady diet of insects, it is probably not necessary to feed them. But you may want to supplement their diet in the hope of growing even more vigorous plants. Or you may be using a terrarium and do not want to hassle with obtaining insect food. Or perhaps you wish to display a plant at an upcoming flower show, and would prefer to present a beautiful but "clean" specimen, without the carcasses of digested insects distracting from the beauty of the leaves. The three numbers associated with fertilizers, such as 30-10-10 or 16-16-16, refers to the percentage of nitrogen, phosphate, and potash contained therein, although some brands also add other minerals and ingredients such as sulfur, blood meal, or iron.

Generally speaking, fertilizers for carnivorous plants are usually best used when diluted to one-fourth or one-half of the manufacturer's directions. It can then be sprayed onto the leaves of the plant until the foliage is wet. It is usually necessary to do this only once or twice monthly during the active growing season. (Do not fertilize dormant plants!) You can also apply fertilizer to the soil of some plants if they have good drainage and are watered frequently from overhead with purified water. Plants grown this way, such as *Nepenthes* in a greenhouse, will have excess minerals leached out of the soil every time the plants are watered. This avoids the much-feared mineral buildup. Otherwise, I would not apply fertilizers to the soil, especially in undrained containers such as bog gardens.

Over time, the excess minerals will build up in the medium and possibly cause harm to the plants.

Some growers prefer to fertilize their plants more frequently using a more heavily diluted ratio. For instance, a 10-percent solution applied weekly to varieties such as *Sarracenia* can work well.

Be warned that by misting or spraying CPs with fertilizers, some will fall on the soil surface and will subsequently encourage algae growth. The same is true of water trays, so it is wise to clean these periodically. Algae growth on soil can be scraped away with a spoon when the buildup becomes substantial, and fresh soil can be laid out to replace it. Mosses will often utilize the small amounts of fertilizers that fall on it from mists or sprays. However, live sphagnum moss dislikes fertilizer buildup and usually succeeds only in pots that are frequently flushed from overhead with pure water—unless one uses Maxsea fertilizer, as mentioned earlier.

It is not wise or necessary to apply fertilizers directly into pitcher leaves. This can upset the delicate chemical balance of its digestive juices and will often produce algae. Simply misting the foliage will benefit the plant enough.

There are many fertilizers and supplements on the market that have been developed for a wide variety of plants and purposes, and different products are available in different countries, but nowhere has one ever been developed specifically for carnivorous plants. Therefore, we must select and experiment with the general forms of fertilizers and supplements available.

Fertilizers for Acid-Loving Plants

These are readily available in most general nurseries. They are used for plants that prefer an acidic soil, such as pines, firs, rhododendrons, and so on. Since many CPs grow in acidic soils, this form of fertilizer is useful. I have used it successfully on all *Sarracenia* species; *Darlingtonia*; most, but not all *Drosera*; temperate, acid-loving *Pinguicula*; most *Utricularia*; and *Dionaea*. Avoid using it on Mexican *Pinguicula* and most *Nepenthes*.

Orchid 30-10-10

This form of plant food is used to promote foliage growth, as opposed to bloom growth. I have used various brands on *Nepenthes, Heliamphora,*

Mexican *Pinguicula*, and most terrestrial *Utricularia*, and I believe it can be used successfully on most of the plants that enjoy acidic conditions, as well.

Epiphytic Fertilizers

You may have to hunt for these in your nursery, as they are specifically developed for epiphytes such as *Tillandsias* and other bromeliads. Many of these plants do not live in soils, and, like carnivores, absorb much of their nutritional requirements through their leaves, often in the form of leaf debris. These fertilizers were developed particularly to enhance mineral absorption through the leaves, and therefore can be quite beneficial to carnivorous plants. I have used brands such as Epiphytes Delight on virtually all carnivorous species, except those few (which I will mention a bit later) that seem to dislike any fertilization regardless of the brand.

Maxsea 16-16-16

An exciting development in recent years has been the invention of the Maxsea brand fertilizer, which was created primarily for marijuana growers in California but has been found to benefit all things botanical. While it contains a standard 16 percent of nitrogen, phosphate, and potash, it also includes several minerals not found in most fertilizers, much of them derived from seaweed.

At California Carnivores, we experimented with Maxsea as a foliar fertilizer on CPs for several months, and found a greatly diluted concentration of 1/2 teaspoon (2.5 ml) per gallon of water (3.8 L) to be ideal. (For "normal" plants the manufacturer recommends a "heaping tablespoon.") This is applied by spraying or sprinkling the plants once or twice a month while the plants are in growth. The seeping of some of this solution into the soil does no apparent harm.

The most astounding result of using this fertilizer was not only the phenomenal growth of our CPs (as well as *Tillandsia* air plants, orchids, and other genera) but also the fact that live sphagnum moss grew profoundly and luxuriously—while all other fertilizers killed live sphagnum. Why this is so remains a total but wonderful mystery.

At California Carnivores, Maxsea is used exclusively on all of our plants, both carnivorous and noncarnivorous alike.

Vitamin B1

Vitamin B1 is a supplement for plants commonly available in nurseries, and it is found in the popular American brand SUPERthrive, which adds other vitamins and ingredients to its solution. These plant supplements are not fertilizers as is commonly supposed, and their use is often controversial among carnivorous plant enthusiasts. Some growers greatly applaud their use, while others find their benefits dubious. Personally, I have found products such as SUPERthrive quite good for carnivorous plants when one follows the manufacturer's instructions. Vitamin B1 has long been used as a root-growth-promoting ingredient, primarily for general garden plants and houseplants. It is most often used when transplanting bare-rooted plants and helps them overcome shock by encouraging root growth. Bare-rooted plants are usually soaked in a solution of ten drops per gallon (3.8 L) of water for half an hour or so before they are put into soil. For general plant care, manufacturers of products such as SUPERthrive recommend a one-drop-per-gallon application with each regular watering. I have noticed that bare-root soaking of carnivorous plants in SUPERthrive or vitamin B1 during the process of transplanting them can reduce losses due to shock. It is particularly helpful when moving tissue-cultured CPs from flask into soil. In my experience, losses due to shock were reduced considerably, if all other conditions were good. Keep in mind that a general application of such products in high doses will increase the growth of algae even more dramatically than is caused by the application of fertilizers. I would therefore not recommend high doses applied directly to soil.

Carnivorous Plants as Pest Controllers

I am often asked whether insect-eating plants are suitable as pest controllers. Sometimes, yes, but more often . . . no.

Carnivorous plants are not the answer to problem insects in the home and garden. They will not zap mosquitoes around your patio, nor devour slugs and snails among your vegetables. They are useless in combating aphids, mealybugs, or scale insects, and in fact are themselves often attacked by such pests.

But in a few limited circumstances, carnivorous plants may be somewhat helpful in controlling bothersome pests. In some situations where African violets, orchids, and other exotics are grown, a few sticky plants such as sundews or Mexican butterworts may be helpful in reducing fungus gnats, a bothersome insect when in its larval stage. Fungus gnats lay eggs on damp soil; these develop into tiny worms that may damage the roots of seedling plants. The gnats are very common among most plants, including cultivated carnivores, but usually are not harmful unless the plants are young and the infestation great. CPs will catch the tiny flying adults in large numbers, providing food for the plant. Terrestrial bladderworts also feed on the larva underground, sucking up the minute worms in their pinhead-sized suction traps. Mexican butterworts and some larger sundews have been known since the Victorian age to be grown in greenhouses as a gnat controller.

Whitefly can be greatly reduced in enclosed places like small greenhouses if a couple of large sundews in hanging pots are grown. An excellent species is *Drosera dichotoma* 'Giant', whose pale, yellowish green leaves attract whitefly.

One hobbyist told me she had great luck catching fleas in her house by using some windowsill- and terrarium-grown CPs. At night she would place large butterworts and cape sundews in saucers on the carpet of the infested rooms. She would place a desk lamp or similar light just above the plants. The heat and light attracted the fleas, and the sticky plants would catch them. I tried this myself one bad flea season and was surprised at the positive result, although it is probably more beneficial to the plant than to any cat or dog!

Probably the most effective CPs to use for insect pest control are American pitcher plants. The trumpet varieties catch many houseflies and wasps, both of which find the *Sarracenia* irresistible. A trumpet plant on a deck, a patio, or a sunny windowsill will catch a surprising number of such pests.

Restaurants with outdoor seating might benefit by having a pot of *Sarracenia* on each table instead of the usual cut flowers. The trumpet varieties can be a very ornamental conversation piece, and the plants will certainly lure and catch many flies. Good species to try would be *S. flava, S. rubra, S. alata*, and *S. leucophylla*, as well as hybrids of these.

Pests and Diseases

Even though carnivorous plants eat insects, there are insect pests that will, unfortunately, eat carnivorous plants. There are also diseases that can attack CPs. To add further to one's anxiety, larger pests such as raccoons, squirrels, blue jays, and small children can wreak havoc on outdoor collections.

Do not despair. First of all, insect pests rarely kill CPs completely and diseases of carnivorous plants are thankfully few. Usually a pest infestation makes itself known to the collector fairly rapidly. Since the plants are so hypnotically beautiful, we can barely keep our eyes off our precious plants for more than a day! Secondly, despite our fears of chemicals, most are safe if one carefully follows the manufacturer's instructions. I strongly recommend isolating treated plants in large, airtight plastic bags or small tanks to make them even safer. I repeat: follow the manufacturer's instructions.

Let me also state something that is quite obvious to the most experienced and confident grower: sometimes plants die. Plants are not immortal. Most can be propagated and passed on generation after generation, but even redwood trees eventually die, even if it takes more than two thousand years! A few carnivorous plants are annuals, meaning they live only a year at most. Many will offer decades of pleasure (I have one *Sarracenia purpurea* that has been in my care for almost forty years). Lastly, even if your cultivation techniques are meticulous, sometimes a plant may kick the bucket for no apparent reason at all. But that is rare. The causes are usually obvious.

That said, let me mention a couple of points. The first is that whenever you obtain a new plant for your collection, inspect it thoroughly—even if you purchased it from a reputable nursery. Try as they might, no nursery is infallible. Secondly, if possible, propagate the plant immediately by taking some leaf or root cuttings. On rare occasions, plants may go into shock and decline in health when moved from one environment to another. Typically they come back with vigor after an adjustment period. For example, I have taken cape sundews and *Nepenthes* from my greenhouse nursery to a windowsill in my home. Occasionally, the existing growing point on the plant stopped developing and went into suspended animation. After many weeks, once the plant had adjusted to the lower humidity and light levels, new shoots appeared at the base of the stems and grew beautifully. I then trimmed off the older tops.

Finally, I must admit that when compared to other plants, there is something a little odd about growing CPs. Because so many of them seem almost animallike—they can appear to have faces and mouths, and most can be hand-fed—for many folks they become more like pets than mere plants! We project our own consciousness onto these plants and become emotionally attached to them. I have seen children weep over the demise of a beloved Venus flytrap whom they affectionately named Liz or Norma. (Note that "whom" seems to fit better than "that" in the previous sentence!)

Pests

APHIDS

These are tiny, sap-sucking insects barely the size of a pinhead. The most common pests of CPs, they attack the newly developing leaves. Since female aphids are born pregnant, they can proliferate faster than rabbits. Aphids can be green, brown, or even black. They are slow moving, and occasionally they fly. Ants like to farm them for their honeydew. Symptoms of an aphid attack are twisted, deformed leaves and sometimes whitish flecks of skin castings around the crown of the plant. Aphids most frequently attack Venus flytraps, sundews, and American pitcher plants, among others.

FUNGUS GNAT

These tiny gnats provide a good food source for plants like butterworts and sundews, and their minute, wormlike larvae are eaten by terrestrial bladderworts. I rarely consider them a pest, as they live in almost any damp soil, but a bad infestation of the larvae in a pot may negatively affect the roots of plants. A spiderlike webbing may be seen on the soil surface, especially after misting.

MEALYBUG

Related to scale, this bothersome pest is the most difficult to eradicate when infestations are bad. Repeated applications of insecticides are usually necessary to eliminate it completely. Mealybugs are slow-moving, soft-bodied, fuzzy little bugs that enjoy attacking pitcher plant rhizomes and leaves. Symptoms of a mealybug infestation are fluffy dabs of cottony

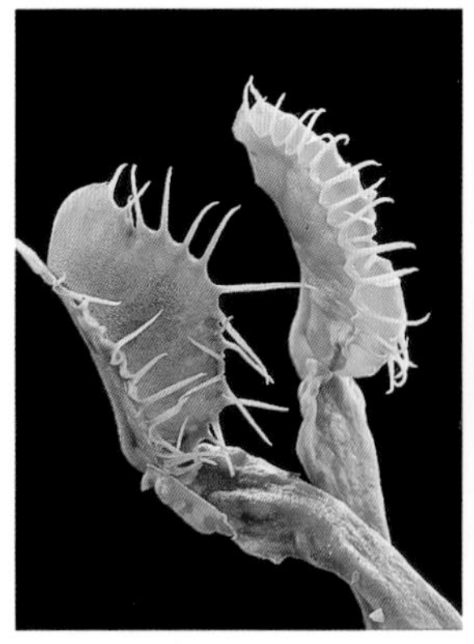

From left to right: Aphid damage on a Venus flytrap. Mealybug on *Nepenthes*. Thrip damage on *Darlingtonia* and *Sarracenia*.

tufts among the growing points and older leaves of the plant. An excess of sooty mold (see page 44) is also a symptom.

RACCOONS, OPOSSUMS, SQUIRRELS, AND BLUE JAYS

These can be an occasional hassle for outdoor plants. These critters enjoy playing with your water trays, carelessly tossing pots around, or digging in bog gardens in search of snails or winter storage caches. Jays love to collect shiny things like jewelry and sundews, and they may also peck at *Sarracenia* to steal its bugs. I have no solutions to these problems, besides waiting for gene-splicing experiments to produce plants large enough to eat them!

SARRACENIA ROOT-BORER AND *EXYRA* MOTHS

These two pests are found only in the southeastern United States, where they are a natural (and exclusive) pest of *Sarracenia*. The root-borer eats rhizomes of the pitcher plant, and is easy to spot due to the reddish orange piles of debris that appear above the rhizome. The moth attacks the pitcher leaves by sealing them up with a webbing at the mouth until the leaf crumples and withers. Both are easily controlled with insecticides.

SCALE

These small, sucking insects live under tiny, protective brown or tan oval shells barely larger than a pinhead. These clamlike shells won't move when you tease them, and they are also farmed by ants. They can be scraped off the leaf, and those killed by insecticides leave their shells intact. They most often attack pitcher plants.

SLUGS AND SNAILS

These occasionally chew a hole or two in newly developing pitcher plant leaves, but after ruining the leaf they seldom eat further, probably due to the taste. They can, however, ruin the appearance of butterworts and sundews, and they love the bromeliad *Catopsis*.

SMALL CHILDREN

If the pitter-patter of little feet is something you hear as you stare hypnotically at your carnivorous plants, and you are sure all of your plants are still in their pots, make sure your CP growing area is child-proof. I will never forget the numbness in a friend's voice when he told me his toddler had pulled the plug on his greenhouse heater the night before a big freeze!

SPIDER MITE

The nightmare of plant lovers, this pest is a problem only in drier climates where humidity and rainfall are low. This minute pest appears as tiny red dots on the leaf, with a faint webbing not unlike that of spiders. I have seen them only on both *Sarracenia* and Venus flytraps in California during the dry summer months. Symptoms of a mite outbreak are vaguely similar to those of thrips (covered next), with a slight discoloration in leaf color and slow decline of the plant. Since mites are not insects but are related to crabs and spiders, insecticides usually have little effect controlling these pests, unless they contain a harmless wetting agent such as canola oil. It's best to use a mitacide such as Avid.

THRIPS

These pests attack pitcher plants, most commonly *Nepenthes, Sarracenia*, and *Darlingtonia*. They are very small, thin black insects that move slowly while eating the surface cells of leaves. Symptoms are a silvery, "scraped" look on the leaf, with small peppery dots of their droppings. Easy to control with good pesticides, this irritating pest is damaging but not necessarily lethal.

Pest Controls

Many insecticides are harmless to carnivorous plants, but others can be rather damaging. Never use soap insecticides, such as Safer Brand Insect Killing Soap, as these products are alkaline and can severely harm your plants. Never use any insecticide from an aerosol spray can; I have found the propellants to be rather harmful.

Listed below are some popular and easy-to-find insecticides that generally do CPs little or no harm. Occasionally, leaves of plants such as sundews may be damaged by prepared insecticides containing petroleum wetting agents, but the plants recover. However, in recent years harmless canola oil has been substituted for petroleum as a sticky wetting agent, with excellent results. Whenever possible, go for wettable powders, as these are mixed with water using no oils. Wettable powders are most often sold through agricultural supply companies.

Some people wonder about natural controls, such as ladybugs (to eat aphids) or mealybug destroyers (beetles that eat, as you might guess, mealybugs). During an aphid outbreak one spring at California Carnivores, we released ten thousand ladybugs to see what would happen. What happened was we found out that ladybugs make an excellent food for carnivorous plants, as all were eaten in about two days. Ladybugs were so drunk on the nectar of pitcher plants, we watched them stepping over aphids as they followed the nectar trails to their death. Members of a gardening group came by (timid growers of African violets, I believe) and, horrified at the wholesale massacre, promptly fled our facilities. The same result occurred when we tried the rather expensive mealybug destroyers.

FLEA COLLARS

Flea collars are a useful insecticide when placed in close proximity to an infected plant (such as in a small tank). Don't let the collar come into contact with water saucers or soils. Use waxy rather than powdery flea collars.

FUNGUS GNAT CONTROLLERS

To kill fungus gnat larvae in soils that can be harmful to seedling CPs, try Summit Chemical's Mosquito Bits. This product looks like chips of charcoal and when mixed with water releases bacteria that will kill the larvae. The chips can be placed in water trays or in a watering can—note that it takes about twenty-four hours to become effective before use.

MITICIDES

Bayer Advanced 3-N-1 Insect, Disease & Mite Control is another excellent pest killer that is harmless to CPs. Avid and Floramite are two superior spider mite controllers.

ORTHENE

An Ortho product, this is the best insecticide to use, as it is systemic and absorbed by the plant, slowly poisoning the insects for many months. Excellent for mealybug, scale, thrips, and aphids. You can order wettable powder Orthene through www.rosemania.com.

PYRETHRINS AND CANOLA OIL

Pyrethrin, an all-natural insecticide produced from plants in the daisy family, has been around for many years; it is safe enough to use on vegetables. More recently it has been mixed with canola oil as a wetting agent. Take Down Garden Spray is one brand available through www.harmonyfarm.com. Another brand is E. B. Stone Rose-N-Flower spray.

SEVIN

This dependable insecticide has been around for decades, commonly available in hardware and nursery supply stores. It can be purchased as a convenient ready-to-use spray, concentrate, and sometimes wettable powder form. It can damage sundew leaves, but the plants recover.

SLUG AND SNAIL POISON

These powders and pellets are effective in keeping these pests under control. Avoid placing them on soils. Water trays and saucers provide good moats to prevent access by these pests. A slug captured with forceps makes a vengeful snack for Venus flytraps. Slightly crushed snails will provide ample vitamins for *Nepenthes*. Yum.

Diseases

BLACK SPOT

This is similar to (or the same as; we're not sure) the disease fungus that attacks plants such as apple trees and roses. I have seen it occur only on Venus flytraps. As its name suggests, black spots appear on the leaves, gradually spreading until all the leaves rot away.

RUST SPOT

This fungus can cause reddish orange spotting on *Nepenthes*. It rarely kills the plant, but it can be unsightly.

BOTRYTIS

This fungus is also known as damping-off disease. Fungus will most often attack newly sprouted seedlings and some rosetted sundews, among other plants. This most often occurs in terrariums that are overly humid, with poor air circulation and low light levels. It can also occur in greenhouses, mostly in winter, when days are cool and overcast. Botrytis appears as a fuzzy gray growth that usually attacks the crown of a plant. I have even seen it in the wild, killing sundews during dank winters. It loves the seeds of *Sarracenia* in late winter, just as they begin to germinate.

POWDERY MILDEW

Similar to botrytis, this fungus can attack pitcher plants such as *Darlingtonia*, mostly in greenhouses, where rain can't keep it under control, resulting in ugly, dark purple blotches on the leaves.

ALGAE SLIME

This can grow in green, oozy puddles along soil surfaces. It is usually harmless to carnivorous plants but is rather gross looking and liable to lose you the blue ribbon at the county fair. Fertilizers and hard water encourage its growth. Scrape it off with a spoon and replace with fresh soil.

SOOTY MOLD

This unsightly pest is actually harmless to CPs. It most often occurs on pitcher plants, where it feeds on nectar. True to its name, it looks like black soot. It can easily be wiped off with a wet towel. However, if sooty mold is produced in excess on *Sarracenia* or *Nepenthes*, it may be also feeding on the honeydew of scale insects and mealybugs, which is a warning sign that you may need to check for these pests.

Disease Control

FUNGICIDES

Physan is an excellent fungicide-algaecide-viracide that will control black spot, rust spot, botrytis, and powdery mildew, and is also used to disinfect medical tools. Serenade is a bacteria-based control, excellent for controlling powdery mildew. Avoid copper-based fungicides, which can kill CPs. However, sulfur-based fungicides seem to do no harm.

PART TWO

WHERE TO GROW CARNIVOROUS PLANTS

Cobra plants and American pitcher plants thriving outdoors in Northern California near the Oregon border, grown by Harry Tryon.

As we have seen, most carnivorous plants come from rather similar habitats. The most pronounced differences are the climates in which the individual species originate. Some varieties may require extremes of heat and cold, wetness or dryness, higher or lower humidity, photoperiod fluctuation, and so on. Therefore, there is no ideal situation in which one can grow all carnivorous plants all of the time. But there are several artificial environments in which a rather large variety can be grown. Some plants, on the other hand, may require rather specialized treatment.

But certainly gone forever are the myths that carnivorous plants can be grown only in humid terrariums or hot, steamy stove houses. Thanks to the years of experimentation by growers around the world, one can find CPs not only in greenhouses and indoor tanks but also on windowsills, on decks and patios, and in corporate offices, outdoor bog gardens, and even basements.

Following is a list of some of the many diverse places you can grow these unusual plants.

Growing Carnivorous Plants Outdoors

Since carnivorous plants are so exotic looking, many people assume they are all from the tropics, which of course is far from the truth. As there is a wide variety of CPs from various climates of the world, chances are you live in an area where at least some, and possibly many, can be grown outdoors. Furthermore, some CPs, such as the popular Venus flytrap, can often grow better outside than inside, even if you live in a climate with cold winters. It is startling to realize that you would probably have better luck growing the flytrap outdoors year-round in New York City than you would if you lived in Key West, Florida.

In Part Three of this book, we will explore the specifics of which species will grow where. Here I will give a basic overview of what carnivorous plant enthusiasts have discovered over the years. This information is, of course, quite general, and most of it from the United States. The terms I use for climate zones are defined at the end of this section.

Unless you live in a swamp, carnivorous plants grown outside will always be container plants. See the sections on Pottery and Containers (page 23) and The Water Tray Method (page 22) for more detailed information. Growing CPs outdoors is not much different from growing the plants elsewhere. The simple rule is that once you have a plant in its proper location, all you have to do is keep it watered well and occasionally weed and trim it.

Many temperate carnivores thrive outdoors in full sun and require chilly to frosty winters for dormancy, such as these American pitcher plants and Venus flytraps. In climates with severe winters, these warm temperature species would need protection.

Direct sun is important to most carnivorous plants. How much direct sun will depend on the plant and where you live. Consider your own climate when deciding where you can grow your plants outdoors. As a rule, if you live in an area of high humidity, full sun will cause most CPs to become colorful and robust. Outdoors in a place like humid North Carolina, American pitcher plants, Venus flytraps, and warm temperate sundews and butterworts (all native to the southeastern United States) thrive in sunny conditions, as witnessed by the attractive bog garden displays at the North Carolina Botanical Garden in Chapel Hill (see photo on page 50).

But if you live in California, where hot summer days often reduce the humidity to 50 percent, 40 percent, 30 percent, and even lower, some protection may be needed to keep up the appearance of your outdoor plants. Morning sun is less harsh than afternoon sun, and screening on a porch or providing a cover of shade cloth will also soften the often burning afternoon rays. Strong sun combined with low humidity will cause nectar burn along the edges of trumpet plants, and it may evaporate the gluey drops of a sundew. Also, the rate of water evaporation from trays and saucers can be startling. In short, if you live in a hot and dry area, CPs will benefit from morning as opposed to afternoon sun, possible protection by screening or shade cloth, and a lot of water. On hot days you may have to fill a deep water bowl up to the surface of a pot and replenish this in a day or two, as the lower humidity will evaporate the water very rapidly.

This may not be the case if you live on the coast in a Mediterranean climate. California, with its numerous microclimates, can be 62°F (17°C) and foggy near the beach—yet just fifteen miles inland, east of the coastal range, temperatures can be in the 90s°F (35°C) and hot and dry.

Certainly the opposite should also be a consideration. A white trumpet plant, *Sarracenia leucophylla*, is adapted to the warm and humid summers of the Gulf Coast, where it is native. If you live in Seattle or right along the Pacific Northwest coastline, *S. leucophylla* may not do very well, as the summers are often cool, foggy, and overcast. You may want to choose a sun-trap area outdoors, such as a courtyard or against a sunny exterior wall. Or better yet, build a small cold frame to house your plants, to allow the greenhouse effect to warm them.

Indoors/Outdoors

Let's say you live in Germany or Wisconsin. You want to have a nice pot of Venus flytraps, cape sundews, or hooded pitcher plants on your sunny patio table. But your winters are too cold to grow them outdoors year-round. No problem.

Venus flytraps and American pitcher plants are dormant and rather unattractive for a few months in winter anyway. They need a chilly winter, but not as cold as Wisconsin's or Germany's. Cape sundews, native to South Africa, can survive a light frost but don't require a dormancy. You can still grow these plants outdoors much of the year simply by moving them indoors for winter. Cape sundews, or other subtropicals that grow year-round without a rest period, can be moved indoors to a sunny, south-facing window when the first frosts threaten in early autumn. Warm temperate plants such as flytraps can be moved indoors before the first severe freezes begin. Just remember that a Venus flytrap requires a cool dormancy. Don't place this plant on a warm, sunny windowsill in winter—you will force it into growth out of season, which will eventually exhaust or even kill it. Instead, choose a cool room, garage windowsill, or porch, out of excessive winter sun, and keep the plant there for its dormancy.

The Cold-Hardiness of Warm Temperate Plants

More experimentation is needed to find out just how cold hardy some CPs really are. But as promising examples, let me mention some extremes I have heard about as a nurseryman and longtime CP grower.

Venus flytraps are native only to the Carolina coastline, yet plants have been introduced and have succeeded for many decades in bogs five hundred miles to the north, in New Jersey.

American pitcher plants such as *Sarracenia flava* and *S. rubra* have been successfully introduced near Vancouver, British Columbia, in Canada, as well as in eastern Pennsylvania. Published accounts indicate that outdoor bog gardens in Vermont, planted with CPs native to the southeastern United States, have survived for twenty years, with mulching for protection in winter.

A customer of my nursery once told me that in Ohio for many years he had an outdoor bog garden planted with southern CPs. Every winter, despite a heavy mulch of hay, the soil of his CP garden rose out of its container like a block of ice, yet all *Sarracenia* and *Dionaea* survived. In such northerly climes, the plants may not grow as fast or produce as many offshoots as in areas of longer summers but still can grow surprisingly well.

A woman in Dallas kept her plants in pots outdoors on a patio. One winter the temperature briefly hit 10°F (-12°C). Her Venus flytraps and American pitcher plants survived, and her cape sundews returned in spring as a result of their thick roots, but she did lose a pot of Australian forked sundews (*Drosera binata*). Even her South African bladderwort (*Utricularia livida*) survived.

Drosera capensis 'Albino' two months after a brief freeze of 24°F (-4°C). New plants are emerging from the older dead stem. Many "subtropical" carnivores are surprisingly hardy if frost levels are not too severe.

Raised bog gardens beautifully display the collection at North Carolina Botanical Garden, Chapel Hill.

Bogs and mini-bogs border a pond at Atlanta Botanical Garden.

Atlanta Botanical Garden has outdoor bog gardens of CPs native to areas somewhat warmer than northern Georgia. Brief lows near 0°F (-18°C) left flytraps, pitcher plants, and other warm temperate species unharmed.

Remember that the duration of deep freezes can be a deciding factor on the cold-hardiness of CPs. Brief drops near 0°F (-18°C) may not kill a Venus flytrap if temperatures quickly rise, but a period of many weeks below freezing may lead to a plant's demise. Also, plants in a bog garden survive freezes better than in exposed pots.

In Part Three on genus cultivation, I will offer more information on the cold-hardiness and heat tolerance of different species of CPs.

Tropical CPs Outdoors

If you live in a tropical climate, you can certainly grow CPs native to similar areas as outdoor container plants. Most growers report that plants such as *Nepenthes* thrive in sunny areas but do best when protected with shade cloth from the hot tropical sun.

Altitude plays an important part in the temperature range of tropical climates. You will have better luck growing lowland *Nepenthes* outdoors if you live in a low-lying tropical region. However, if you live on a mountain above 3,000 feet (914 m), where night temperatures can cool considerably, highland *Nepenthes* will be a better choice than the lowland varieties.

Growing warm temperate plants in highland tropical climates is also possible. People living in Hawaii, on the edge of the tropical zone, can often succeed with Venus flytraps and *Sarracenia* outdoors if they live in a cooler, high-altitude region. Folks in the lowlands may have to bare-root such plants and store them over winter in the refrigerator.

Subtropical climates, such as Miami's, and warm temperate Mediterranean ones, such as San Diego's, rarely see frost. Yet these climates differ in another way: the latter city will experience cool nights year-round, even when summer days are warm. Highland *Nepenthes* thrive in such weather and can be successful outdoors, but only near the humid coast. Yet only lowlanders would survive the hot summer nights of Miami. In both places, it is wise to bring the plants indoors when a cold snap is predicted. Tropical *Nepenthes* may slow down or stop growing during the cooler winter months, but they usually become vigorous as the weather warms up.

There are a few *Nepenthes* that have been found to be surprisingly cold hardy. I have found that *N. khasiana*, one of the most cold-tolerant tropical pitcher plants known, can tolerate brief temperature drops into the 30s°F (3°C) without damage when protected overhead from frost.

Some tropical sundews are also frost hardy. *Drosera adelae* has been known to return from its roots after light freezes in the San Francisco area.

Bog Gardens

A carnivorous plant bog garden can be a center point of drama and intrigue as well as beauty. Artificial bogs can be large or small. They can be placed in the ground to appear natural, or constructed in a container as a dish garden for the deck or patio. Even greenhouses can be the home of such a savage garden. Provided you have enough pure water to keep the bog wet, its maintenance can be simple for many years, requiring only basic weeding and trimming.

The plants you can grow in an outdoor bog garden are primarily the same plants your climate allows you to grow outdoors year-round. Greenhouse bogs will naturally offer protection in harsher weather, and smaller containers or mini-bogs can be moved around should protection be needed. Even in cold temperate climates, larger bogs that are in the ground can be protected from extreme temperatures by mulching during winter dormancy, allowing plants to be grown far from their native habitats. There is even a method for growing individual carnivores in a mixed, noncarnivorous garden setting, as in a rock garden or one consisting primarily of cacti and succulents.

Setting up a bog garden outdoors is not unlike constructing a small pond, but it's much simpler. Instead of water, the container is filled with a wet mixture of peat and sand. There are two construction methods. The first is to dig a hole conforming to the dimensions you want for your bog. Its depth should be a minimum of about 10 to 12 inches (25–30.5 cm). Deeper bogs will hold more water, so they may require less watering in drier climates. The bog's width should leave the center easily accessible from the edge. However, a wider bog can be equipped with stepping stones to make access to the center easier. Remember, you won't be able to walk in your bog to weed or trim plants later.

After digging, the shallow hole is lined with sheet plastic. It is best to use pool or pond liners designed to hold water. They are much more durable than greenhouse sheet plastic. If the ground soil is very rocky, you may want to fill in a thin layer of sand or dried peat over the dimensions of the hole, so that small, sharp rocks or pebbles won't pierce the liner when it is weighted down with the bog's soil. A liner that overlaps the bog's edge can be trimmed and later covered and decorated with rocks.

You should not construct a bog in depressed areas of your property where there is a risk of flooding during heavy rains. I recommend punching several holes in the liner around its periphery, around 2 to 3 inches (5–7.5 cm) below the surface of the bog's soil. This will allow drainage of excess water after heavy rains, so plants won't be permanently waterlogged.

The second method for setting up a bog garden is somewhat simpler. Instead of a pool liner, one can use a prefabricated container such as a children's wading pool or a molded plastic pool designed for a small pond or water garden. Water gardens have become so popular in recent years that there are many intriguing shapes, sizes, and designs available at nurseries and garden centers. Simply excavate a hole to accommodate the molded pool and set it into the ground. Again, I recommend drilling a few drainage holes into the sides below the soil surface.

The peat and sand should be of equal parts, and premixed with water before you add the mixture to the bog container. Be sure to pack it firmly. I also like to vary the depth of the surface, so that you have well-drained mounds and wetter low areas to better suit the individual species' preference. It is often fun to have a shallow depression in the bog, where water will be permanent; you can grow aquatic bladderworts or *Aldrovanda* there. In a very large bog you may want to have a shallow moat of water encircling the whole garden, to bar access of certain crawling pests such as slugs and snails. Aquatic carnivores or other water plants can be grown in the moat.

An outdoor bog garden at California Carnivores.

Once the soil is in the bog garden and packed firmly, planting can begin. If you use potted plants, a bog garden can be planted at any time of the year. If you obtain bare-root plants for your bog, it is best to set this up in the late winter or early spring as the plants are coming out of their dormancy.

Designing the layout of your bog will take foresight. If your garden is to be viewed from all sides, it is best to group taller plants, such as trumpet plants, toward the center. Smaller varieties, such as Venus flytraps, parrot pitcher plants, and sundews, are best along the outer edges. If you locate your bog so it is viewed primarily from one side, such as against a fence, taller plants would naturally look best toward the back, gradually terracing the smaller varieties toward the front. Keep in mind the species' growing habits. A single crown of a sweet trumpet plant (*Sarracenia rubra*) can spread into a large dense mass in a few years. *Drosera binata*, likewise, can spread through its root system. You might consider growing such plants in a porous container within the bog, such as peat pots or cloth bags, but these in time will deteriorate anyway, and most people prefer their plants to spread.

The species of plants you can grow in a bog garden of this type are those found naturally in a peat-based soil. You could not grow *Drosophyllum* in such a bog, for example. Most enthusiasts enjoy growing a wide variety of carnivorous plants in such a garden, such as *Sarracenia, Dionaea, Drosera, Pinguicula, Cephalotus, Byblis,* and *Utricularia*. But equally stunning are theme bogs, such as a garden of all trumpet plants, either all one species or mixed, including hybrids. Alternatively, a low-growing bog can be just as impressive, using species such as *Sarracenia purpurea* and *S. psittacina, Dionaea,* rosetted *Drosera*, and temperate *Pinguicula*.

If you live in Connecticut, yet want to grow warm temperate species from the Florida Panhandle or elsewhere, you can protect the bog in winter. The fact that it is set in the ground will automatically offer some protection, but you should take further precautions: after the autumn equinox, as the plants go dormant and the first hard frosts occur, trim off their old leaves so that most of the taller foliage is removed. Then scatter several inches (10 cm) of a mulch that is later easily removed, such as pine needles, hay, or layers of burlap. A blanket of snow will insulate the bog even further.

After winter, when the last of the severe frost danger is over, the coverings can be removed. Some plants that do not form true winter resting buds, such as the short-lived *Drosera capillaris* or *Pinguicula lusitianica*, will probably have died off. Such varieties usually return from, or can be reintroduced by, seed from the previous season.

NONCARNIVOROUS ADDITIONS

Noncarnivorous plants can make attractive additions to the savage garden. You can check with your local nursery or water garden specialist for wet-tolerant plants to add to your bog. Ornamental grasses, in particular, make handsome contrasts. The following are some suggestions. Most of these plants are native to the damp pinelands of the southeastern United States.

- Yellow-eyed grass (*Xyris baldwiniana*)
- Pipewort (*Eriocaulon compressum*)
- Shoe button (*Syngonanthus flavidulus*)
- Bog buttons (*Lachnocaulon anceps*)
- Grass pink (*Calopogon tuberosus*)
- Rose pogonia (*Pogonia ophioglossoides*)
- Yellow star grass (*Hypoxis hirsuta*)
- Coneflower (*Rudbeckia graminifolia*)
- Comfort root (*Hibiscus aculeatus*)
- Bay blue-flag (*Iris tridenta*)
- Japanese blood grass (*Imperata cylindrica* 'Rubra')
- Teena turner grass (*Isolepis cernuus*)
- *Disa* orchids from South Africa

Carnivores in a Mixed Garden

Carnivorous plants growing among cacti and succulents, ornamental grasses, or in a rock garden can be rather startling. Yet this is easily accomplished if you grow carnivorous plants in undrained containers that are set into the ground. I like to use large plastic pots that have no drainage holes, such as the ones that floral shops use to store cut flowers. Plant your carnivore in such a pot with its preferred soil mix, then dig a hole in your garden to accommodate the container and set it into the ground. The rim can be disguised with rocks or mosses. Make sure you water the CPs frequently and separately from your other garden plants. Either avoid using sprinklers on the CPs if your tap water is hard, or protect them from the sprinkler by covering them with small patches of sheet plastic when you water.

Long, rectangular containers set in the ground along a sidewalk can be planted with rows of *Sarracenia* trumpets for an unusual and effective border. Circular plastic garden bowls planted with a mass of cape sundews, Venus flytraps, or flowering terrestrial bladderworts can be the centerpiece of a mixed "alpine" rock garden. If your climate permits it, a drained container of climbing *Nepenthes* can be set in the ground among small palms and ferns for an unusual tropical effect.

The Mini-Bog

Miniature carnivorous plant bog gardens are one of the most popular and simple ways to grow CPs. These are set up in the same way as larger, in-the-ground artificial bogs. The only difference is that you'll use smaller, freestanding containers. These can be placed on sunny decks, patios, or balconies. If you live in a climate with harsh winters, the containers can be moved to a bright garage window or basement window for the winter to accommodate the bog's dormancy, if needed, or into a sunroom if the plants grow year-round.

Containers for mini-bogs? You can use a wide variety of plastic garden bowls for this purpose, or even wine barrels or wooden planter boxes that are lined with sheet plastic under the soil (remember, the container needs to be undrained). Wooden window boxes lined with plastic and planted with forked sundews or pitcher plants certainly draw more attention than geraniums, and snare houseflies before they come through the window!

Jana Olson Drobinski's intriguing mini-bog in an antique tub.

A visit to your local weekend flea market can often turn up fabulous and unusual mini-bog containers. Undrained, glazed ceramic flower pots, urns, bonsai dishes, salad bowls, large ashtrays, water basins, and cut-flower vases are just some of the possible containers in which to grow a mini-bog. The smaller containers can be ideal for tiny bogs of miniature plants: pygmy and rosetted sundews, short-rooted butterworts, terrestrial bladderworts—all do well in shallow containers. Larger and deeper containers can be planted with a similar variety of CPs as a sizable in-the-ground bog garden, but on a smaller scale.

For many years I grew the following in a 14-inch (35.5 cm) plastic garden bowl: *Sarracenia rubra* ssp. *wherryi, S. flava*, and *S. psittacina*; *Drosera capensis, D. spatulata, D. filiformis* ssp. *tracyi,* and *D. patens* × *occidentalis*; *Utricularia livida*; plus a clump of yellow-eyed grass.

An interesting variation on the mini-bog is the miniature island bog. In this setup, I use a plastic garden bowl with a removable plug, which I discard. I cover the hole with a handful of long-fibered sphagnum to prevent the leakage of soil. Then I fill the container with a peat-and-sand mix and plant it as I would any other mini-bog. I then set this container in a similar but much larger garden bowl, leaving the plug intact. This larger container acts as the water tray, and I keep the water fairly deep most of the time. Not only will this method allow for less watering stress, but also the moat around the mini-bog will act as a barrier to slugs and snails. Further, the moat becomes the ideal place to grow aquatic carnivores such as *Utricularia gibba* or *U. purpurea*, or the waterwheel plant, *Aldrovanda*.

CLIMATE ZONE TERMS

The following terms referring to climate zones will be found throughout this book.

COLD TEMPERATE: Warm summers, with very cold and long winters, with temperatures below freezing lasting for many days at a time.

TEMPERATE: Warm summers, with winters having many cold snaps, usually of brief duration and not below 22°F (-5°C), although colder temperatures may rarely occur.

WARM TEMPERATE: Warm summers, with occasional cold snaps of short duration usually not below 28°F (-2°C), although colder temperatures may rarely occur.

SUBTROPICAL: Warm summers with mild winters. Occasional cold snaps, but rarely below freezing, or 32°F (0°C).

TROPICAL: Few temperature extremes, but no freezing temperatures expected. Highland tropical means warm days and cool nights. Lowland tropical means hot days and warm nights.

MEDITERRANEAN: Warm, dry summers and cool, wet winters. May be temperate, warm temperate, or subtropical. Cool Mediterranean climates have cool but dry summers.

Windowsill Growing

A *Nepenthes* in your living room? Sundews over your kitchen sink? Butterworts in your bathroom?

A few decades ago, many carnivorous plant enthusiasts would have raised an eyebrow at such possibilities, assuming that to grow a CP indoors, terrariums would be a requirement. Not necessarily so. From London to New York to San Francisco to Melbourne, many carnivores are finding happy homes in people's houses and offices, requiring minimal care while offering maximum pleasure.

This does not mean you can grow a Venus flytrap wherever you grow a parlor palm. Most houseplants are cultivated for their low light tolerance, and in fact may dislike or be burned by direct sun. CPs in general are sun lovers, and this is perhaps the most important fact to

Nepenthes khasiana, *Drosera capensis* 'Albino', and *Pinguicula moranensis* growing in the author's living room.

consider when choosing carnivores to grow indoors.

I use the term "windowsill" to stress the point. Most carnivorous plants, to be successful indoors, usually need to be as close to a window as possible. A windowsill or tabletop right next to the glass is the brightest place in a room, and this is what CPs typically require.

Furthermore, since most carnivores are sun lovers, direct sunshine streaming through the window for at least part of the day is also a necessity, although there are a few exceptions. Most CPs need direct sun for a minimum of two to five hours during the growing season. Bear in mind that the sun moves about in the sky from season to season, so there are few windows that receive the same amount of sun throughout the year. Many people in the northern hemisphere assume south-facing windows are the sunniest, but this may be true only in the winter when the sun is low in the southern sky. Come summer (the primary growing season for most plants), when the sun moves directly overhead, a southern-exposed window may receive no direct sun at all, especially if your house has a roof overhang.

Most growers have found that east- or southeast-facing windows—which in the northern hemisphere receive cooler morning sun—are probably the best indoor location for CPs. West- and southwest-facing windows are also quite good, but if the afternoon sun is too hot in your house, screening or sheer curtains may be needed to make the sunlight less harsh and keep the plants from burning.

There are several factors to consider when growing carnivorous plants indoors. If you wish to grow temperate plants, such as the Venus flytrap or American pitcher plants, you will have to take careful heed of their dormancy requirements. It is the shortening of the daylight period that will trigger the plant's rest period. Dormancy also requires cooler temperatures. Your plants won't necessarily require temperatures as cold as in their native habitat, but dormant plants should certainly be protected from hot sun and warm temperatures indoors.

It is therefore wise to move such dormant plants outdoors, if your climate can sustain them, or perhaps to the coolest north-facing windowsill in a room that is not overly heated, especially at night. Basement and garage windowsills can fill this requirement, as can enclosed porches that get chilly in the winter.

Some folks who have no such environment to place dormant plants may instead remove the plant from its soil late in the season and store the rhizome in an airtight plastic bag after trimming whatever leaves may be remaining on the plant. A few strands of damp, long-fibered sphagnum moss or a handful of moist peat moss can be added to the bag, and the bag may then be refrigerated over winter. Come late winter, the rhizome is repotted and the plant returned to the windowsill.

Temperate plants should enter their dormant phase sometime during mid- to late autumn, and usually begin active growth as the daylight period increases in late winter or early spring. In the United States, I like to remind people that plants should enter a rest period sometime between Halloween and Thanksgiving and can be brought out of dormancy as early as Valentine's Day but no later than Easter. By March, and the approach of spring, the well-rested plants will begin the season's growth.

Of course, many subtropical and warm temperate CPs will continue to grow through winter and won't require a cold rest period, although they may slow down their growth. These plants, like *Cephalotus*, cape sundews, or the Mexican butterworts, will be happy year-round on the windowsill. You may want to grow such plants on a sunny, south-facing window in winter, and by spring, move the plants to an east- or west-facing window for the summer.

Alternatively, you may grow plants outdoors in the summer and then move them to a sunny windowsill for winter, to protect them from unsuitably cold weather. In a city like Boston, you might grow Mexican butterworts or cape sundews outdoors for the summer, then move them to a windowsill before the first frosts. Remember, if you do this with a Venus flytrap or yellow trumpet plant (perhaps because your winters are too cold to leave them outdoors), make sure you respect their rest period and keep them in a cool and sunless window.

Success with windowsill growing may depend on where and how you live. Humidity is certainly important for carnivorous plants indoors, but it not as important as good light. If you live near a coastline, indoor humidity is often suitable for CPs. If you live in a desert community, air-conditioning or evaporative coolers will often help plants grown on windowsills. In winter, heating your house may drop the humidity drastically, especially with energy sources such as woodstoves. You may want to keep a kettle filled on your woodstove to replenish water vapor.

Misting indoor CPs can be quite beneficial, and I recommend you keep a spray bottle of purified water near your plants if your house is on the dry side. Wetting the foliage in the morning and evening is appropriate if your humidity is low.

The following are all suitable for the windowsill, if conditions are appropriate: Venus flytraps; most Mexican butterworts; *Sarracenia* species and hybrids, although all will require much direct sun; most terrestrial and epiphytic bladderworts that are subtropical to tropical in origin, such as *Utricularia livida, U. sandersonii, U. reniformis*, and *U. humboltii;* aquatics such as *U. gibba; Cephalotus; Byblis* species; sundews such as rosetted subtropicals, cape sundews, *Drosera regia,* and *D. binata*; *Darlingtonia* if your house is cool; and highland *Nepenthes* such as *N. khasiana, N.* × *rokko, N. alata*, and *N. ventricosa*, including hybrids. Some lowland hybrids can also do surprisingly well, such as the beautiful *N.* × *dyeriana*, which has some highland ancestry.

Terrariums and Tanks

The method of growing carnivorous plants in tanks is still one of the most popular and enjoyable ways to raise carnivores. Not only can a well-presented tank of flesh-eating plants rival a commonplace aquarium for decorative beauty, but also the maintenance can be but a couple of hours a month or less—and, unlike fish hobbyists, you won't have to feed the inhabitants every day! Also, tanks and terrariums can be kept almost anywhere in the home, school, or office. I strongly suggest you consider cultivating the plants under grow lights, as recommended in the Artificial Light section (page 27).

There are several ways of growing CPs in tanks. Although the traditional terrarium may be the first that comes to mind, this old-fashioned, soil-at-the-bottom-of-an-aquarium style may not necessarily be the best.

The following section deals with a few basic terrarium styles, moving from the easiest to the most difficult to maintain.

Greenhouse-Style Terrarium

This is my favorite method of growing CPs in tanks. Basically, you take an empty aquarium, set fluorescent grow lights along its glass-covered top, and grow the plants in pots that sit in individual water saucers.

There are several reasons why this method is superior. The first is variety. Kept in individual pots and saucers, *Nepenthes* can grow in their preferred soil mix, while rainbow plants can grow alongside in a completely different medium. You can grow temperate plants with the tropicals most of the year, but can easily remove a Venus flytrap or purple pitcher plant during the winter and place it elsewhere for its dormancy. Should aphids suddenly appear on a sundew recently added to the tank, it can be promptly removed and treated before the pest spreads to other plants.

Keeping the potted plants in individual saucers allows you to maintain the wetter/drier cycle that some plants, such as Mexican butterworts, may require. Also, species such as *Byblis liniflora, Cephalotus*, and the *Nepenthes* do not appreciate waterlogged conditions all of the time.

A greenhouse-style terrarium under grow lights in the author's home. None of these plants require a winter dormancy: subtropical sundews, Mexican butterworts, highland tropical pitcher plants, sun pitchers, bladderworts, and rainbow plants do quite well at room temperature year-round.

Probably the biggest relief comes at cleaning time. Large and heavy tanks can be an ordeal to clean when the algae and splashed soil particles become unsightly. Using the saucer method makes cleaning the tank much easier.

A SUBTROPICAL GREENHOUSE-STYLE TANK

Plants should be kept in individual pots; on a twelve- to fourteen-hour photoperiod; minimum temperature 55°F to 65°F (13–18°C); maximum temperature 75°F to 85°F (24–29°C). The following plants do well:

Highland *Nepenthes* varieties, cape sundews, and rosetted subtropical sundews, such as *Drosera aliciae, D. spatulata, D. hamiltonii, D. venusta,* and *D. anglica* 'Hawaii'. Tropical sundews such as *D. adelae, D. schizandra, D. prolifera,* and *D. intermedia* 'Cuba'. Forked sundews like *D. extrema* and *D.* 'Marston Dragon'. Most pygmy sundews. All Mexican, subtropical, and warm temperate butterworts. *Byblis liniflora*. *Heliamphora* species and hybrids. *Cephalotus*. Subtropical terrestrial bladderworts, such as *Utricularia sandersonii, U. livida,* and *U. calycifida*. Tropical epiphytic bladderworts such as *U. longifolia, U. alpina,* and *U. reniformis*.

The listed plants do not require cold dormancies and most will generally be attractive year-round. It is best to reduce the photoperiod to ten hours in winter, gradually increasing the period to between twelve and fourteen hours in summer. This will usually keep flowering times on schedule and also trigger winter rosettes in the Mexican butterworts and gemmae production in pygmy sundews. If your tank contains only plants that originate near the equator, like highland *Nepenthes*, a year-round photoperiod of twelve to fourteen hours is best.

A HEATED, GREENHOUSE-STYLE TROPICAL TANK

When warmed to a minimum of 65°F to 75°F (18–24°C), the following plants do well; the photoperiod should be twelve to fourteen hours. Lowland *Nepenthes* varieties, Mexican butterworts, *Genlisea, Byblis liniflora*. Tropical sundews such as *Drosera adelae, D. schizandra, D. prolifera, D. petiolaris,* and *D. indica* varieties. Tropical epiphytic bladderworts such as *Utricularia longifolia, U. alpina,* and *U. reniformis*. Tropical terrestrial bladderworts such as *U. pubescens* and *U. calycifida*.

The third benefit of the saucer method for the greenhouse terrarium is that lower-growing plants like rosetted sundews can be raised closer to the lights by placing the pot and saucer on an empty, upside-down pot or some other pedestal. Further, plants with larger drooping or pendulous leaves, such as forked sundews or *Nepenthes*, will be shown to better advantage with the pedestal method. Your basic square or circular green plastic pots work well here, or you may choose an opposite approach and grow the plants in a variety of ceramic, glazed pottery.

The Potted-Landscaped Terrarium

This method is rather similar to the preceding one except that the space between the pots is filled with fine orchid bark, lava rock, perlite, pumice, or mosses to give the appearance that the plants are planted in soil. Long-fibered sphagnum makes a good medium to use for this method, and live sphagnum growing along the surface can be rather attractive. Trimming the live moss will be necessary to prevent it from overgrowing some of the shorter potted plants. One can also use orchid bark, pumice, lava rock, or perlite as a base to hide the pots, with sphagnum as a top-dressing, although the whiteness of perlite may be distracting. You can still keep the plants in individual saucers, but it is easier to set them on a base of moss, pumice, lava rock, or bark, and then raise or lower the base to suit the plant's wetness or dryness requirements. The water table would be visible through the glass.

It can be fun to decorate a tank with this method. Raised pots of *Nepenthes* can be hidden with Spanish or sphagnum moss draped over the exterior. Potted bog grass, orchids, or ferns can make the tank more natural-looking. I like to set mossy branches, rocks, and *Tillandsia* air plants on the soil surface of such a tank, giving the appearance of a tropical jungle even if the plants are not native to such an environment. Although lethal as a growing medium, decorative or green sheet mosses can be used as a soil dressing as well.

When the plants are kept in individual pots, you have the advantage of moving them around, as with the greenhouse-style tank. Cleaning out and redoing the whole tank is necessary, but usually only every one or two years.

The Classic Terrarium

Despite its limitations, the classic terrarium, where plants are grown in soil at the bottom of an aquarium, is still a feasible way of growing some CPs. One drawback is that variety may be limited, since some plants may require a dormancy while others do not. Furthermore, setting up and redoing a terrarium can be a sloppy ordeal, and if one plant succumbs to disease or pests, the whole terrarium may soon follow suit. Finally, some of the plants you may wish to grow in your terrarium may require somewhat different cultivating techniques from the others. A Mexican butterwort, for example, needs a somewhat drier winter and different soil than *Cephalotus* does.

The most important thing to remember is that your selection of plants should be of varieties that share a similar soil and climate in their wild habitat. In other words, don't mix temperate Venus flytraps and tropical pitcher plants, since not only do they come from differing climate zones, but their soil requirements are rather different as well.

A habitat terrarium is fun to create and maintain. For this type of tank you choose plants that grow in the wild together, such as a southeastern U.S. savanna. Typically, this is easy to set up: on the bottom of the terrarium, lay 2 or 3 inches (5–7.5 cm) of horticultural sand, pumice, or lava rock. This allows good drainage and a visible water table. On top of this, place a layer of premixed and wetted peat moss and sand; about a fifty-fifty ratio is good, or use 80 percent peat moss to 20 percent perlite. Sloping the medium is a wise idea, so the water-loving plants can be placed in the lower portions and better-drained plants in the upper or higher grade of the medium. Usually, 3 to 4 inches (7.5–10 cm) of soil is good for most plants.

In a temperate terrarium such as this, when grown under artificial light, remember to fluctuate the photoperiod between winter and summer as discussed in the Natural Light section (page 25). Also, dormancy will be critical in winter (see the Dormancy section, page 26); the tank should be moved to a cooler room, outdoors to a covered porch or patio, or to a garage window.

Permanent indoor terrariums, grown under artificial lights on a twelve- to fourteen-hour photoperiod, are more suitable to tropical and

TEMPERATE CLASSICAL TERRARIUM

Use a soil base of half peat to half sand or 80 percent peat to 20 percent perlite. Photoperiod should be reduced to eight hours in winter, with cool temperatures: 40°F to 60°F (4–16°C). By summer, increase the photoperiod to sixteen hours with temperatures of 60°F to 90°F (16–32°C). The following plants do well: Venus flytraps, low-growing American pitcher plants such as *Sarracenia purpurea, S. psittacina, S. rubra,* and the smaller hybrids. Warm temperate butterworts such as *Pinguicula caerulea, P. pumila, P. lusitanicum, P. primuliflora,* and *P. lutea.* Terrestrial bladderworts such as *Utricularia subulata* and *U. cornuta.* Temperate sundews such as *Drosera rotundifolia, D. intermedia, D. capillaris,* and *D. filiformis.*

As mentioned, this type of terrarium is difficult to maintain due to the often chilly to cold temperatures the plants require during their winter dormancy or rest period. Many subtropical plants will also succeed in such a tank if temperatures do not drop below freezing.

subtropical CPs. In a similar peat/sand soil, you can grow cape and rosetted sundews, *Cephalotus*, some Mexican butterworts, and rainbow plants, although the latter three varieties will need good drainage. You could also scoop out some of the peat and sand, replacing it with a pocket of *Nepenthes* soil for better drainage, and try tropical pitcher plants there. Some *Nepenthes*, such as *N. mirabilis*, often tolerate swampy conditions in the wild.

A word of warning, however. Some plants in a classic terrarium may run wild if left unchecked. *Drosera capensis* will produce so much seed that if the flower stalks are not removed, cape sundews will come up in the hundreds within a year. Some terrestrial bladderworts, like *Utricularia livida*, will spread in months throughout the whole tank until it is a mass of lovely flowers—but you may not see much else in your terrarium! Trimming off the flowers will do these plants no harm.

Heating a Terrarium

Should you desire to grow lowland tropical CPs in a tank, it is best to supply a heat source so the minimum temperature is maintained at roughly 70°F (21°C). To accomplish this, visit your pet shop. A simple way to heat a terrarium is to place an aquarium heater in a large jar of water and set this in the tank (the jar can be disguised with decorative mosses). Set the thermostat between 80°F and 90°F (27–32°C). The warm water will also provide additional humidity.

Alternatively, various heating pads and "hot rocks" can be used to warm a greenhouse-style terrarium, particularly those designed for reptile habitats. The pads can be placed under the tank. The thermostat should be set at around 70°F (21°C) minimum.

Cooling a Terrarium

If you are growing highland plants that prefer chilly nights, such as highland *Nepenthes* or sun pitchers, one way to drop temperatures is to place refreezable ice packs in the tank when you go to bed at night, removing and refreezing the packs the following morning.

Maintaining a Terrarium

There are four ways to make your terrarium more attractive and easier to maintain. One is to attach your grow lights to a timer, so they go on and off without your having to be around. The second is to keep your tank ventilated. This means having an air gap of 1 to 2 inches (2.5–5 cm) along the top of the tank to allow good air circulation. A constantly steamed-up tank with an overabundance of humidity and stagnant air is a sure invitation to mold and fungus. A third important suggestion is to line the back and sides of the tank with a reflective material such as Mylar, white cardboard, or mirrors. This will greatly enhance the strength of light on the plants and color them up beautifully. Some growers place a removable reflector on the front of the tank, removing it when they are home or wish to view the plants. This will cause the light to bounce around the tank, and the vivid colors of the plants—even some distance from the grow lights—will take your breath away. Finally, I like to keep a spray bottle of purified water near the tank. Giving the terrarium a heavy mist in the morning and evening increases the humidity and circulates the air.

Greenhouses

The controlled environment of a greenhouse is no doubt the ideal place to grow beautiful carnivorous plants. From my own lengthy experience with the joys and horrors of greenhouses, I strongly recommend purchasing a commercial, prefabricated house from one of the many professional suppliers. An Internet search will provide many vendors. In the United States, the reputable company www.charleysgreenhouse.com is the "candy store" of greenhouse suppliers.

Greenhouse Conditions

What follows is a listing of the basic concerns for a good and functional greenhouse; for more detail I suggest you also consult one of the many good books available at libraries, bookstores, and home-supply businesses.

GLAZING AND FRAMING

Glazing means the clear or translucent material that the roof and walls of a greenhouse are made of, usually attached to a metal, wood, or PVC plastic frame.

Plastic PVC pipes covered with polyethelene plastic sheeting is suitable—if you enjoy rebuilding your greenhouse after every windstorm! I don't recommend these types of greenhouses at all.

Forget old-fashioned fiberglass as well. It cracks, light transmission is reduced over time, and it offers no heat retention. Also forget the granddaddy of glazing: glass. It shatters too easily, especially in severe hailstorms and earthquakes.

In this twenty-first century, the foolproof glazing of greenhouses is polycarbonate. This amazing plastic invention is lightweight and durable, and light transmission can be excellent for more than twenty years. It is also nearly impossible to break.

Polycarbonate glazing comes in three types: single, double, and triple wall. The latter two offer "insulation" to retain heat through the inclusion of air gaps between the sheets of polycarbonate—just like double-glazed glass windows for homes. When California Carnivores relocated our facilities in 2002, we used triple wall to construct

California Carnivores' warm-house display room

80 percent of our new greenhouse, especially the roof. We were amazed that when the temperatures outside dropped to 28°F (-2°C), inside temperatures dropped no lower than 48°F (9°C) without heating. Fuel costs for heating during cold weather were drastically reduced.

FLOORS

While some greenhouse owners may prefer laying an expensive concrete floor, I have found water-permeable, plastic woven ground or weed cloth used by landscapers, covered with about 4 inches (10 cm) of pea gravel, to make an excellent floor. The cloth deters deep-rooted weeds from taking hold, and when wetted, the gravel retains moisture to increase humidity.

HEATING

There are certainly some carnivorous plants that can be grown in unheated greenhouses, even if you live in a climate with cold winters. But most people will find it necessary to heat their greenhouses if they wish to grow a wide variety of CPs. It is always best to have the heat mechanically controlled with thermostats, and fans are helpful to circulate the air while the heat is on. In a small greenhouse you can often make do with space heaters designed for this specific use, and many larger heating systems are available from greenhouse specialists.

SHADING

This is a requirement for most greenhouses, and it is very helpful to cut down on heat buildup from the sun (the well-known greenhouse effect). The two most common shading methods are shade cloth and whitewash. Usually a 50-percent shade cloth is ideal for CPs. Any denser cloth, such as 70 percent, will be too darkening for the health of most CPs. Whitewash is usually gypsum mixed with water, which is then "painted" on the roof and sides of the house. However, whitewash frequently has to be replaced, as rain will gradually wash it away. And be warned that gypsum, being alkaline, is lethal to most CPs if it is applied by accident to their soil or water. Therefore, shade cloth is often the safest product to use, unless your greenhouse is made of polycarbonate. Shade cloth can be harmful to polycarbonate, so in that case, use a whitewash instead. Greenhouse suppliers also sell shading compounds that can be diluted to a preferred strength, and also come in varying degrees of durability and rain resistance.

Highland *Nepenthes* in California Carnivores' warm house.

FANS

Exhaust fans are another requirement for the greenhouse; they should be set by a thermostat to turn on when heat reaches a certain temperature, usually between 70°F and 80°F (21–27°C). Automatic vents on the opposite end of the greenhouse will allow cooler outside air to enter.

SWAMP OR EVAPORATIVE COOLERS

This is an often necessary method to cool and humidify most greenhouses. As the temperature goes up, the humidity goes down, and on a very hot, dry day, simple exhaust fans will draw hot and dry air through the house. There are many types of swamp coolers available on the market. The general systems have water dripping through pads, which is recirculated from a water reservoir. Fans are necessary to produce the effect. In small, freestanding units that look like air conditioners, a built-in fan blows air through the pads, cooling and moistening the air through an "evaporative" method. These are often called "personal coolers" in department stores. The method is the same for larger units, but typically a wall on one end of the greenhouse contains the large, water-moist pads, while the fans are usually exhaust fans located on the opposite wall of the house. When the fans are engaged, hot, dry air from the outside is drawn in through the pads, cooling the temperature while adding humidity by evaporating water.

MISTING SYSTEMS

These cooling systems operate by atomizing water into a cooling mist, and they can be quite effective for many greenhouses, especially where low humidity is a concern. But be warned that the water source should be purified or of a low mineral content. Otherwise, the mist nozzles will become clogged by mineral salts and the CPs will be harmed by the minerals deposited on their leaves and soil. (And they will look unsightly when these dissolved salts dry up on their foliage.) Some plants, such as sundews, will not appreciate having their leaves frequently wetted by mist, but others, such as *Nepenthes*, thrive in such conditions.

HUMIDIFIERS

Small room humidifiers can be purchased through greenhouse suppliers as well as from home-supply businesses and drugstores. These small units are inexpensive and helpful when added humidity is needed in small greenhouses. Typically, humidifiers have small reservoirs to hold water, which is then heated and released as a vapor. These units can also be helpful for larger propagation chambers.

AIR CONDITIONERS

If you live in a climate with extended periods of warm summer nights, yet you wish to grow carnivorous plants adapted to chilly night temperatures, you may consider installing an air conditioner in your greenhouse. Some varieties of CPs require a substantial drop in night temperatures for good growth, among them highland *Nepenthes, Darlingtonia, Heliamphora,* and *Drosophyllum*.

Types of Greenhouses

Here I will slightly amend the general information on greenhouse types and simplify their descriptions to suit the cultivation of carnivorous plants.

THE COLD FRAME

This type of greenhouse is unheated, except through solar radiation (the heat of the sun). This means its usefulness is dependent primarily on the climate you live in. Thus in areas that experience cold winters, where temperatures are often below freezing, in a cold frame you can grow only carnivorous plants that come from similar climates. Generally these species have long, sustained winter dormancy. Some suitable species are *Pinguicula* from the northern latitudes, such as *P. vulgaris, P. macroceras, P. longifolia*, and *P. grandiflora; Sarracenia purpurea* ssp. *purpurea*; *Drosera rotundifolia, D. anglica*, and *D. intermedia* from the northern latitudes; and those species of *Utricularia* native to cold temperate climates.

Alternatively, if you live in a warm temperate climate where cold winter temperatures only occasionally drop below freezing, you can draw from a wider selection of cold frames. For example, an unheated greenhouse in the southeastern U.S. coastal plain, much of California, the Mediterranean countries, or the subtropical coastal areas of Australia can be home to a wide range of CPs. All *Sarracenia* would be suitable, as would *Dionaea*. Temperate to hardy subtropical sundews, butterworts, and bladderworts will also do well, as will the dewy pine and cobra plant, among others.

THE COLD HOUSE

This type of greenhouse is heated when temperatures drop below freezing (32°F, 0°C). Suitable plants for this environment are all *Sarracenia, Dionaea, Darlingtonia, Drosophyllum*, temperate and subtropical *Drosera, Pinguicula, Utricularia*, and *Aldrovanda*. The pygmy and tuberous sundews from Western Australia would be at some risk if the foliage were allowed to freeze.

THE COOL HOUSE

Heated at 40°F (4°C), a cool house is a safer environment for the plants just suggested for the cold house. In addition, you would be able to grow some Mexican butterworts, *Cephalotus*, and at least one *Nepenthes* species: *N. khasiana*. A few of the more cold-hardy highland *Nepenthes* species and hybrids may stop growing in winter but otherwise should survive. *Heliamphora* would do well also.

THE WARM HOUSE

Heated at around 50°F (10°C), this type of greenhouse is ideally suited for growing a wide variety of carnivorous plants. Although a low of 50°F (10°C) is considerably warmer than many CPs would experience in the wild during their winter dormancy, the naturally shortened photoperiod of winter would keep them dormant for a suitable enough length of time,

This hobby warm house in California, owned by Damon Collingsworth of California Carnivores, gets cool summer nights, appropriate for highland *Nepenthes* and *Darlingtonia*. Where summer nights are warm, air-conditioning from midnight to dawn may be necessary.

unless you live in the tropics. However, the warm house would not be the best for plants from a northern latitude, such as *Pinguicula vulgaris* or *Drosera rotundifolia*, which may do better in winter if they're placed under benches where cooler temperatures might be maintained. It is also not wise to allow the greenhouse to get too hot on sunny winter afternoons; when the interior temperature reaches around 60°F (16°C), exhaust fans should be set to cool the greenhouse.

Varieties suitable for the warm house are *Sarracenia*; *Dionaea*; most temperate to subtropical *Drosera,* including the pygmies and winter-growing species; all subtropical and Mexican *Pinguicula*; *Darlingtonia*; *Heliamphora*; *Roridula*; temperate, subtropical, and epiphytic *Utricularia*; *Drosophyllum*; *Byblis*; *Cephalotus*; *Aldrovanda*; *Ibicella*; the carnivorous bromeliads; and the highland *Nepenthes.*

In short, the only unsuitable varieties for the warm house are the species from the most northern latitudes and the lowland tropicals, particularly lowland *Nepenthes.*

THE HOTHOUSE

The heat goes on in the hothouse when the temperature drops below 60°F (16°C). This is a good environment only for those plants considered tropical. Highland *Nepenthes* do well here, particularly if daytime temperatures don't exceed 85°F (29°C). However, much more suitable are the lowland *Nepenthes*, although a few of these may slow their growth in winter. Tropical sundews, *Genlisea, Byblis liniflora, Ibicella,* and Mexican butterworts thrive in hothouse conditions. Humidity should be high at all times.

THE STOVE HOUSE

Temperatures in the stove house are maintained above 70°F (21°C), with hot days in the 80s°F and 90s°F (27–35°C). Only true lowland tropical species should be grown here, such as the lowland *Nepenthes, Genlisea,* tropical *Byblis*, and *Drosera*. High humidity is also essential.

Remember that if your greenhouse is large enough, you can always build a small enclosure within it and heat this section separately from the main growing area. An example of this would be heating the main section at 50°F (10°C) (warm-house conditions) while heating the enclosure at, let's say, 65°F (18°C) (somewhere between hot- and stove-house requirements). You can then grow lowland tropicals in the enclosure while more temperate or highland tropicals are maintained in the main house. The heating can be accomplished with small space heaters set on a separate thermostat. Heating pads used for propagation (there are many types available) are also suitable for the warmer enclosure. Set at 70°F (21°C), a heating pad will keep the pots warm while the air may be somewhat cooler.

Likewise, smaller enclosures within the greenhouse can be cooled separately with an air conditioner or small swamp coolers, should this be necessary in your area to grow the plants requiring cooler climates.

BOG GARDENS IN THE GREENHOUSE

Although the most common way to grow CPs in the greenhouse is to set potted plants in water trays or saucers, a bog garden is certainly an option. You could, if you wish, put a bog garden into the ground of your greenhouse as you would outdoors. But it is much easier to construct a tableau or boxed bog garden right on the benches or staging of the greenhouse.

A fourteen-inch (35.5 cm) mini-bog of South African sundews and bladderworts.

You could use a prefabricated plastic container such as a wading pool and set this on your greenhouse bench, but it is often cheaper to construct your own. A simple way is to make a rectangular wooden frame out of 8 × 2-inch (20 × 5 cm) boards. I like to use a sheet of plywood as the base. This box is then lined with a sturdy, 6-mm sheet plastic, stapled along the interior. Simply fill the box with your soil mix of prewetted peat moss and sand, and landscape as you would any outdoor bog garden. Depending on the size of your boxed container, features such as *Utricularia* puddles or moats can also be added, just as with outdoor gardens.

Tabletop tableau and mini-bog gardens at California Carnivores.

PART THREE

THE PLANTS AND HOW TO GROW THEM

Carnivorous plants have evolved into a wildly diverse variety of designs. Left to right, back row: *Heliamphora heterodoxa* x *ionasi*, *Sarracenia leucophylla* x *rubra* ssp. *alabamensis*, *Drosera filiformis* 'California Sunset'. Middle row: *Drosera venusta*, *Utricularia livida*, *Darlingtonia californica*, *Cephalotus follicularis*. Front row: *Dionaea* 'Gremlin', *Pinguicula agnata* x *moranensis*, *Nepenthes spathulata* x *hamata*.

The Venus Flytrap

(Dionaea muscipula)

The Venus flytrap.

Awe inspiring when first seen in action, the Venus flytrap is without doubt the most famous of all carnivorous plants. In the early 1760s, then-governor of North Carolina Arthur Dobbs brought public attention to the plant for the first time, calling it the "Fly Trap Sensitive." A few years later, specimens were sent to England, where it was the first plant ever suspected of being carnivorous. The origin of the plant's name has a controversial history. Naturalist John Ellis first described it and named it after the Greek goddess Dione, one of the many mythological references to Venus. The species name *muscipula* literally means "mousetrap." The Swedish botanist Carolus Linnaeus, who invented the binomial system of Latin names to identify all species of life on earth, called it a "miracle of nature." But he was unconvinced the plant was truly carnivorous, because it reversed the laws of nature as he believed God had intended. However, most other botanists were soon to disagree.

Venus flytraps are native only to the coastal plain of southeastern North Carolina and extreme northeastern South Carolina, in roughly a hundred-mile (161 km) radius from Cape Fear, near the town of Wilmington. Once very abundant, the species is now rather threatened, due primarily to habitat destruction and to a lesser extent collection for the retail nursery trade. This latter practice is changing, thanks to a growing reliance on nursery tissue culture.

A pot of Venus flytraps grown on the author's sunny porch.

Unfortunately, the loss of habitat is occurring at an alarming rate, due to a rapidly growing population and the draining of wetlands for lumber, agriculture, and residential development. Native populations also decline because of the prevention of naturally occurring brushfires, usually caused by frequent lightning strikes. Flytraps are choked out by thick scrub that would otherwise be burned back by these fires. Although the Venus flytrap has been introduced with some success in states such as New Jersey and California, it has been naturalized in great abundance in only a small area of the Florida Panhandle.

In their native Carolina habitats, Venus flytraps grow primarily in sandy, peaty soils in damp areas on the edges of swamps, fens, and pocosins (upland swamps). Typically they are found in open, sunny, wet savannas or grasslands amid sparsely scattered pines. Their climate is warm temperate and humid. In the summer, the days are hot and nights are warm. Winters are chilly, with occasional periods of frost, but only rarely does it snow, and hard freezes are brief.

Venus flytraps are perennial plants. Grown from seed, they usually take about four to five years to reach maturity and can live for many decades.

Mature plants produce a rosette of leaves averaging 4 to 8 inches (10–20 cm) in diameter. The leaves consist of two parts: the petiole (actually an expanded leaf base) and the trap, which is the true leaf. Adult plants usually have traps averaging 1 inch (2.5 cm) long. The leaves come up from a short, thick rhizome or underground stem. The few thick, black roots are several inches (15 cm) long.

In late winter or spring, at the start of their growth for the season, the plants produce a small rosette of leaves with wide, heart-shaped petioles, which usually hug the ground, and small traps.

The plants usually bloom in spring, sending up a wiry stem 8 to 12 inches (20–30.5 cm) tall. Each of the several white flowers are about 1 inch (2.5 cm) across. Flowering can have an exhausting effect on plants in cultivation, so unless you want seed, the flower stalk is best clipped off when it's 2 or 3 inches (5–7.5 cm) high.

The flowers of the Venus flytrap. Now that you've seen them, cut them off!

After flowering, larger traps are produced as summer approaches. The plants continue to produce leaves throughout summer. Older leaves, several weeks old, turn black and die. These should be trimmed off. As autumn arrives in late September, all of the summer leaves die away and are replaced by smaller, low-growing traps.

In mild winters some of these traps remain, but they react indifferently to the capture of insects. During hard frosts, all of the leaves may die off.

The trapping mechanism of the Venus flytrap is amazing; it prompted Charles Darwin to call the plant "one of the most wonderful in the world." It was he who ultimately offered proof of its carnivorous nature. In his book *Insectivorous Plants*, Darwin performed many experiments on the flytrap, including some strange ones: he found he could paralyze the trap by making certain incisions on it, and that the traps could be anesthetized with ether.

The exact mechanism of the trap is still a mystery that is hotly debated. The general principle is as follows: The trap consists of two halves not unlike a clamshell. The outer margins are lined with teeth, or cilia. When the trap is open, the two halves or lobes are actually concave or dished inward. Each lobe has three or four tiny trigger hairs set near the center.

A sweet nectar is produced by glands found along the inner base of the teeth that rim the trap. Insects, most often ants or flies, are lured by this nectar to enter the trap. As the insect moves about, drinking the nectar, it needs to touch or bend two of the tiny trigger hairs or one hair twice within twenty seconds to spring the trap. What occurs next is startling.

A mild electrical current runs through the trap. The cells on the outer walls of the lobes lengthen in less than a second. This quick growth causes the concave, dished shape of the lobe to rapidly reverse itself. The trap snaps shut, causing the teeth to intermesh—imprisoning the insect in a cage.

The trap does not close tightly right away. Darwin surmised that this allowed small insects to escape through the intermeshed teeth, so the plant wouldn't waste time and energy eating an insignificant meal. But if a larger insect is caught, its struggling will stimulate the trigger hairs even further. In a few hours, the lobes are pressed tightly together and the trap seals itself. Glands on the inner surface of the lobes begin to secrete digestive juices. Shortly the insect drowns in this fluid.

It takes a flytrap from four to ten days to digest its prey. The soft parts of the insect are dissolved and this fluid is absorbed by the plant. When the trap reopens, only the dried, shriveled exoskeleton of the

An earwig . . . seconds from doom . . . Gotcha!

insect remains. Rain may wash the carcass out of the trap. More often, spiders are lured by the crusty shell and become a second meal. In the wild, large ants and spiders are the most frequent prey.

If a trap is closed empty, as a result of a falling leaf or rude finger, it will usually reopen within one or two days. Each trap can catch one to three meals, after which the trap and petiole die and turn black. Depending on the age of the leaf or the size of its prey, a trap may turn black after one meal. This is normal. Remember to trim all dead traps off the plant.

Varieties and Forms

Although there is only one species of *Dionaea muscipula*, making it a monotypic genus, a few natural variations in the form have been found in the wild. The most common and typical have small, rosetted traps held close to the ground in autumn, winter, and spring, whereas in summer the traps are held a few inches (10–15 cm) in the air by narrow, elongated petioles. The interior of the traps is invariably pink to reddish. However, there have been forms found in the wild that lack any red coloration at all, remaining green even when grown in full sun. Others may have deeply red interior traps rather than the more common pinkish red. Some forms never seem to produce taller summer traps at all; they are larger during the summer but remain close to the ground or rosetted. Remember that to enhance any coloration of a Venus flytrap, full, direct sun or adequate artificial lighting is necessary during the growing season.

Cultivars

A cultivar is a "cultivated variety" of a plant grown by human beings. Cultivars are chosen and named for distinctive or unusual characteristics the plant may have. A cultivar may have originated in the species' native habitat, but more often these are bred by growers who cross-pollinate individual plants, hoping for unusual qualities to appear in the offspring. However, with Venus flytraps, the vast majority of cultivars originated in tissue-cultured plants that mutated due to the chemicals and hormones used to sterilize and propagate the plants in test tubes. Not only can the leaves be mutated, but sometimes the flowers can be, too.

In the 1990s, there were only a handful of Venus flytrap cultivars known. Presently there are many hundreds, although only several dozen have been officially registered with the International Carnivorous Plant Society (ICPS). For a complete list of known cultivars, one can go to the Society's website at www.carnivorousplants.org, which also has links to photos and histories of the individually named plants. Here I will mention some of the more popular or curious cultivars, some of which are in wide circulation.

All-Red Venus Flytraps

Dionaea 'Akai Ryu' or 'Red Dragon'

The first mass-produced tissue-cultured flytrap cultivar ever produced. Bred by Ron Gagliardo at the Atlanta Botanical Garden, this clone is entirely colored a deep dark maroon, except for the flowers. It can reach enormous size as well, with traps sometimes surpassing 1 1/2 inches (3.8 cm) in length, on summer petioles 6 or 7 inches (15–18 cm) in height. The English translation of its Japanese name is most often used by growers.

Dionaea 'Green Dragon', a sibling of 'Red Dragon' produced by the Atlanta Botanical Garden.

Dionaea 'Green Dragon'

A sibling of 'Red Dragon'. It isn't green, as one might suppose; appearing nearly identical to 'Red Dragon' in size and coloration, it has a distinctive greenish band on the outside of the trap along the base of the cilia teeth.

Dionaea 'Royal Red' and 'Clayton's Red Sunset'

Two popular all-red flytraps produced in Australia, the latter from Triffid Park Nursery.

Dionaea 'Petite Dragon'

Of the countless varieties of maroon or red flytraps, one curious form is 'Petite Dragon'. In the 1990s, Bob Zeimer, an editor of the *Carnivorous Plant Newsletter* (*CPN*), was visiting California Carnivores when the first shipment of young, tissue-cultured 'Red Dragon' plants arrived. One that he took with him matured into a tiny all-red plant with miniature leaves and flower stalks that forked.

Dionaea 'Dente'.

The Dentate Group

There are many flytraps that have very short cilia teeth, resembling those of a shark.

Dionaea 'Dente'

A handsome clone that produces medium-size summer traps on short, upright petioles, the interior of the trap often turning a deep red.

Dionaea 'Red Piranha'

A cloned hybrid of 'Red Dragon' × 'Dente'. It is entirely red with short, jagged teeth.

Dionaea 'Sawtooth'

A typically colored flytrap whose cilia have been reduced to very short, irregularly pointed and jagged teeth.

All-Green Flytraps

Dionaea 'Justina Davis'

There are a number of Venus flytrap clones that lack any red coloration at all. Notable among these is 'Justina Davis', named by *CPN* editor

Barry Rice. The name commemorates the child bride of the discoverer of the Venus flytrap, Governor Arthur Dobbs, who was in his sixties at the time of their marriage.

Dionaea **'Gremlin'**
This cultivar, developed at California Carnivores, is a vigorous, clumping, all-green clone.

Giant Flytraps

While typical Venus flytraps have traps averaging about 1 inch (2.5 cm) or slightly more in length, a number of clones, mostly produced through breeding, have traps considerably larger, approaching or surpassing 2 inches (5 cm) in length.

Dionaea **'B-52'**
A formidable clone produced by Henning Von Schmeling of the United States. The cultivar name doesn't signify the giant jet bomber but happened to be a code for plants he was breeding. Vigorous and impressive, these large specimens also can clump, producing a mass of large, yawning traps.

Dionaea **'Ginormous'**
Similar to 'B-52', this cultivar is a gigantic-enormous clone registered by California Carnivores.

A typical mature plant on the left, *Dionaea* 'B-52' on the right.

One interesting observation of these extra-large flytraps is that quite often when triggered, the traps don't close correctly, as in normal-size plants. One lobe may warp closed, but the other may not for two or three seconds, and sometimes the intended prey may escape. This malfunction of a larger trap may be why Venus flytraps haven't evolved naturally into mouse-traps or rat-traps or even larger.

Hideously Mutated Flytraps

It is interesting to observe that with the hundreds of random, accidental mutations occurring in tissue-cultured flytraps, none has really improved on the naturally evolved design. Some, in fact, are grotesque and reduce the trap to nonfunctional. Since some of these clones find it difficult to catch and eat prey, they should be foliarly fertilized to make up for their dietary loss.

Dionaea 'Cupped Trap'

This mutation produces traps that have their lobes joined at the outer edge, preventing the trap from closing around the prey.

Dionaea 'Wacky Traps'

Nearly impossible to describe, this clone has traps once said to imitate cartoon character Bart Simpson's hair. Their irregularly jagged, cilia-free lobes seem almost painful to look at.

Dionaea 'Fused Tooth', a tissue-cultured mutation grown as a novelty.

Dionaea 'Fondue'

Registered by Guillaume Bily of France, this indescribable mutation may produce normal traps in spring but the summer traps look like molten waxy flytraps that have melted into hideous shapes. *Fondue* means "molten" in French.

Dionaea 'Fused Tooth'

Some of the cilia on this oddity are fused together.

CULTIVATION

(See Parts One and Two for further information.)

SOIL RECIPE: Flytraps thrive in a mix of one part sand to one part peat. Another good mix is four parts peat to one part perlite.

CONTAINERS: Best in plastic pots or glazed ceramics. Use 4- to 5-inch (10–12.5 cm) pots for single, mature plants. A grouping of several plants looks good in 6- to 8-inch (15–20 cm) pots. They do well in deeper mini-bogs and bog gardens.

WATERING: Use the tray method, keeping the soil damp to wet year-round. Flytraps do not appreciate persistent waterlogged conditions and do best with a lower water table.

LIGHT: Full to part sun.

CLIMATE: Warm temperate plants, flytraps need warm summers and chilly winters. Tolerant of light frost and brief freezes.

DORMANCY: Venus flytraps require a chilly winter dormancy, a period when they slow down and usually stop growing.

OUTDOORS: Venus flytraps do very well in temperate, warm temperate, and Mediterranean climates.

BOG GARDENS: Venus flytraps do very well in temperate, warm temperate, and Mediterranean climates. Mulch in colder zones.

WINDOWSILLS: Good candidates for sunny windowsills. Best kept cooler for winter dormancy.

TERRARIUMS: Seasonal candidates for the greenhouse-style tank, best removed in winter. Good for the classic, temperate terrarium.

GREENHOUSES: Venus flytraps do well in cold houses, cool houses, and warm houses, and in cold frames in warm temperate climates.

continued

Cultivation, *continued*

FEEDING: Readily accept houseflies, large ants, spiders, sow bugs, or pill bugs. Moistened dried insects are also accepted.

FERTILIZING: Outdoor plants usually catch an abundance of insects and rarely require any fertilizer. However, windowsill, terrarium, and greenhouse plants can benefit from a foliar spray of Maxsea or other acceptable fertilizer, applied once or twice a month during the growing season. Avoid saturating the soil with any fertilizers.

TRANSPLANTING: Potted flytraps are invigorated when transplanted into fresh medium every one to two years; best done in late winter, while in a dormant state. If flytraps are bare-rooted and transplanted during the growing season, they may undergo shock but usually slowly recover.

GROOMING: As older traps and leaves die off and turn black, they are best removed using scissors. Never try pulling off recently blackened leaves, as they remain attached to the rhizome for a while and you can uproot the entire plant.

A baby Venus flytrap about two years old.

PESTS AND DISEASES: Aphids are the most common pest; the result is twisted and deformed new leaves. They are effectively controlled by insecticides like pyrethrin/canola oil or others recommended in Part One in the Pests and Diseases section. Flea collars placed very close to the plant or in an enclosed plastic bag or terrarium work well also. Use the waxy type flea collars, not the powdery.

In hot, dry climates, spider mites can attack flytraps. Use a mitacide such as Avid.

Black spot fungus can appear on plants in an overly wet and humid environment. Use a sulfur-based fungicide for control.

PROPAGATION

SEED: The individual flowers open for a few days. The anthers release the pollen immediately. The stigma is receptive in two or three days, when it appears fuzzy. Transfer the yellow pollen grains from anthers to a receptive stigma. Rubbing open flowers together will usually pollinate them. Numerous small, shiny black seeds are produced in about six weeks.

The seed can be sowed immediately or refrigerated for later use. Scatter the seed on their preferred soil mix. Keep humid and in bright light. Germination occurs in a few weeks. Transplant the seedlings when they are about one year old.

DIVISION: Occasionally Venus flytraps will produce offshoots and develop into a clumping plant. These are best divided in late winter to early summer. Make sure that each crown of leaves has a separate root system before you divide them.

LEAF CUTTINGS: In spring or early summer, peel leaves off the rhizome with a downward tug, making sure the whitish base of the leaf is intact. Lay the whole leaf right-side-up on a peat-sand mix or long-fibered or milled sphagnum. Lightly cover the base of the leaf with a pinch of soil. Keep humid and damp in bright light. Clear plastic bags or seed propagation trays

continued

work well to ensure high humidity. In a few weeks plantlets will appear at the leaf base or on the leaf margins. In a few months, when the plantlets have several leaves and roots, they can be potted up individually. Leaf cuttings can produce mature plants in about two years.

FLOWER STALK CUTTINGS: If you cut off the flower stalk of a flytrap to allow larger spring leaves to develop sooner, you can remove the stalk when about 3 inches (7.5 cm) high and use it to propagate another new plant. Trim off the small flower buds and lay the stalk on peat moss, slightly covering the cut ends with a little peat, similar to propagating leaf cuttings. One or more plantlets usually develop from the tips of the cut flower buds in several weeks.

TISSUE CULTURE: Flytraps can be propagated in vitro through seed, newly emerging leaves, and flower buds.

The American Pitcher Plants

(Sarracenia)

A French print of *Sarracenia.*

Beautiful and easy to grow, American pitcher plants may be the most ravenous and underappreciated plants in horticulture. The bizarre and often handsome leaves may catch thousands of bothersome insects such as ants, flies, and wasps. Their flowers are showy, brilliant, and very unusual—a wonderful bonus to an already handsome class of foliage plants. Yet of the few varieties sometimes available on the mass market, tens of thousands die needlessly due to poor handling in countless nurseries, where the plants often go down the dismal path to the compost heap, like millions of Venus flytraps before them. What a shame! The *Sarracenia* are one of the simplest carnivorous plants to grow, and certainly among the most fun and rewarding.

There are currently assumed to be eight species, all of which are confined to the southeastern United States, with the exception of one species extending north along the seaboard and into the Upper Midwest and much of Canada. For such unusual and once-common plants, they were slow to be recognized by the early European settlers. The first published illustration was of *S. minor* from Florida in 1576. In 1700, Tournefort described *S. purpurea* from plants sent to him by Dr. M. S. Sarrazin of Quebec, and Linnaeus followed his lead, naming the genus *Sarracenia* in 1731. William Bartram, in a 1793 book about his travels in the southern United States, first mentioned the vast quantity of insects caught in the pitcher leaves but doubted that the plants could benefit from them. Darwin suspected their carnivorous nature but did not study them. Many botanists at this time were studying the genus, but it was mostly the work of Dr. Joseph J. Mellichamp in the 1870s and 1880s that ultimately proved *Sarracenia* eat insects. The general knowledge of the plants was greatly extended by the field and laboratory studies of Dr. Edgar Wherry in the 1930s, and more recently by Dr. Donald Schnell and Frederick Case. There is still some controversy over the natural species and subspecies status of the genus. Here I will follow the general conclusions of Schnell.

The typical habitat of American pitcher plants is found on the southeastern coastal plain of North America. Scattered individuals or dense colonies are most frequently found in permanently wet, open, grassy savannas, fens, swamps, and similar wetlands. The soils are various, ranging from low-nutrient sand—often so fine it resembles clay—to peaty soils derived from sphagnum mosses. Stands of long-leaf pine or other trees may populate the area, but the pitcher plants will prefer the sunniest areas and avoid the dense shade of trees. In their natural state, these wetlands were frequently the targets of lightning strikes, and the ensuing brush fires kept scrub, bushes, and tree seedlings in check, keeping the habitat grassy and open. In the past, Native Americans started fires for a similar purpose, primarily to maintain open fields for the ease of hunting deer. The pitcher plants thrived in such areas.

A wet meadow of white trumpets in the Florida Panhandle. Scenes like this were once abundantly common.

Once abundant throughout much of the southeastern coastal plain of North America, *Sarracenia* populations have become vastly reduced, mostly due to the drainage of wetlands for pulp wood production, shopping centers, and housing developments. Also, the suppression of naturally occurring fires can eventually choke out the pitcher plants with the spread of brush and trees.

At least one species (*S. oreophila*) and a subspecies (*S. rubra* ssp. *jonesii*) are severely endangered plants in mountain or foothill remnant wetlands above the coastal plain, in places such as northern Alabama and the North and South Carolina piedmont foothills of the Appalachian Mountains. The northern purple pitcher plant, *S. purpurea* ssp. *purpurea*, is the only species found north of Virginia. Its habitat is primarily wet, acidic sphagnum bogs found in scattered areas of northeastern North America and throughout much of Canada. Oddly enough, in the Great Lakes area this plant is also found in wet, marly, alkaline wetlands.

The climate of the southeastern coastal plain is considered warm temperate. Rain falls throughout much of the year, and summers are warm and humid. Winters are cool and often frosty at night. Occasional brief deep freezes and rarer light snowfalls also occur. In its northern range, *S. purpurea* ssp. *purpurea* experiences extremely frigid winter conditions, often with a lot of snow.

The seeds of *Sarracenia* are usually dispersed in autumn. In late winter and spring of the following year, they begin to germinate as the weather warms. By the end of the first year's growth the seedlings will have tiny pitcher leaves 1 or 2 inches (2.5–5 cm) long. A typical plant takes around five to eight years to reach maturity. Over the years, individual plants of most but not all species develop a thick, branching underground rhizome, from which several growing points emerge.

The annual growth cycle of mature plants begins after winter dormancy. Most plants start the season by flowering. Each growing point will send up one bloom on a 1- to 3-foot-high (0.3–0.9 m) stalk. The first pitcher leaves appear from the rhizome soon after the developing flower bud, but the plant flowers first, before any pitchers open. They wouldn't want to eat their pollinators!

The flowers are showy—depending on the species, they may be from 1 to more than 4 inches (2.5–10 cm) in diameter. The flowers hang upside down from the tall stem. The five petals, from 1 to 4 inches (2.5–10 cm) in length, hang pendulously from the bloom. Four of the eight species have yellowish petals; the other four have various shades of red petals. The flowers are rather unusual and beautiful, and most are in bloom from one to two weeks. Bees are the primary pollinators. Each species flowers at slightly different times between late February and May. In areas where two or more species grow together, this usually eliminates cross-pollination of species. However, quite often two or more species may flower at similar times, often resulting in swarms of hybrids. Hybrids of *Sarracenia* species occur fairly frequently in the wild. At the turn of the century, many commonly found hybrids were thought to be species.

The flower of *Sarracenia purpurea* ssp. *venosa* var. *burkii*.

Almost all of the species have scented flowers. The aroma may be strong or mild, sweet to musty. As the petals drop off, the ovary and style remain all summer as the seed develops for autumn release. The fruit or seedpods are often so attractive in their own right that many people assume them to be flowers even though they have lost their colorful petals. In some species, the pods slowly tilt upward so that during the autumn the seed won't become trapped in the upside-down umbrella-shaped style of the seedpod.

It is after petal drop that the first pitcher leaves of the season open for business. The differences in the structure and trapping mechanisms of the various species are rather dramatic and will be described later. But generally speaking, insects are lured to the leaves by a combination of nectar and color. It is assumed that a drug in the nectar strongly assists in the trapping of prey. A drug called coniine has been isolated from the nectar of *S. flava*. This narcotic causes paralysis and eventually death to those insects drinking enough of it. While most American pitcher plants catch insects in a pitfall method, wherein the prey fall into tubular leaves from which they cannot escape, at least one, *S. psittacina*, catches victims with a one-way trap, and another, *S. purpurea*, drowns its prey.

Pitcher leaves are produced from spring until late summer or early autumn. Some species produce leaves more or less on a continuing basis throughout the growing season. Others will send up their leaves in crops: spring, early summer, and late summer. The individual leaves are in prime condition for a period of time ranging from several weeks to a couple of months. After this period they begin to deteriorate, often filled with insects. After the autumn equinox most species stop leaf production and by winter are in a dormant state. Usually most leaves brown and decompose over winter, although some species may hold on to some of their leaves during this time, only to lose them rapidly when spring growth resumes, as with *S. purpurea*.

Insects that encounter the purple pitcher plant, *S. purpurea*, drown in collected rainwater, where they slowly decompose by bacterial action and weak enzymes. All of the other species trap their prey in tubular leaves, near the bottom of which digestive acids and enzymes are produced and secreted more heavily as more insects are caught. Microorganisms also play a part in digestion. The soft parts of the insects break down, and the plant slowly absorbs this nutritious soup, gaining nitrogen, potassium, phosphorous, and other trace elements that are lacking in the plant's soils. Research indicates these minerals play a significant role in the plant's ability to flower and set seed.

The Purple Pitcher Plant

(SARRACENIA PURPUREA)

Sarracenia purpurea has the widest range of any American pitcher plant; it is divided into two subspecies, with several varieties and forms.

In all of the plants the pitchers are decumbent, more or less growing close to the ground in a rosette fashion. The leaves can be up to 6 to 8 inches (15–20 cm) in length. The hollow leaves resemble colorful, flared cornucopias. A large, often undulating collar is open to the sky, allowing rainwater to be collected by the leaf, unlike the leaves of other *Sarracenia*. This collar is also covered in bristly, downward-pointing hairs. Insects often cling to and slip from these hairs, which are wet with nectar. The prey tumble into the water below, where they drown.

Subspecies

Sarracenia purpurea ssp. *purpurea*

The subspecies known as *S. purpurea* ssp. *purpurea* is sometimes called the northern pitcher plant. Its range is throughout much of Canada, the Great Lakes region, and the eastern seaboard from Labrador, Canada, and south to New Jersey, where it meets its southern subspecies. It has also been introduced and naturalized in some regions of Europe and New Zealand and a few sites from northern California to Washington State. The pitchers are narrower than those of its plumper sister to the south and are often more numerous and more densely packed. They frequently last through the most frigid winters.

Normally a denizen of acid sphagnum bogs, *S. purpurea* ssp. *purpurea* can also be found in alkaline fens around the Great Lakes, where the smaller pitchers take on a more brittle consistency. These plants revert to normal when moved to acid conditions. The flowers have red to purplish petals, and the plants may bloom in midsummer in its most northerly range.

Sarracenia purpurea ssp. *purpurea* form *heterophylla*

The form *heterophylla* is an uncommon strain in which the all-red color is absent—instead, the pitchers and flowers are entirely yellowish green.

Sarracenia purpurea* ssp. *venosa

This is the southern subspecies of the plant, and the one most commonly found in cultivation. Its range begins in the New Jersey Pine Barrens, where intermediates between it and *S. purpurea* ssp. *purpurea* occur. From New Jersey, the plants grow south along the coastal plain to Georgia. Their range continues west (with a gap in central Georgia) across the Florida Panhandle, southern Alabama, and Mississippi.

Sarracenia purpurea ssp. *venosa*.

Varieties and Forms

Sarracenia purpurea* ssp. *venosa* var. *venosa

This variety is found south of New Jersery, primarily along the coastal plain of the Carolinas. The pitchers can be rather plump and fat, from green with much red veining to nearly entirely plum red. The collar is broad and rather undulating. The pitchers can reach up to 7 inches (18 cm) in length and have fine soft hairs along the exterior. The flower petals are red on the outside and paler inside.

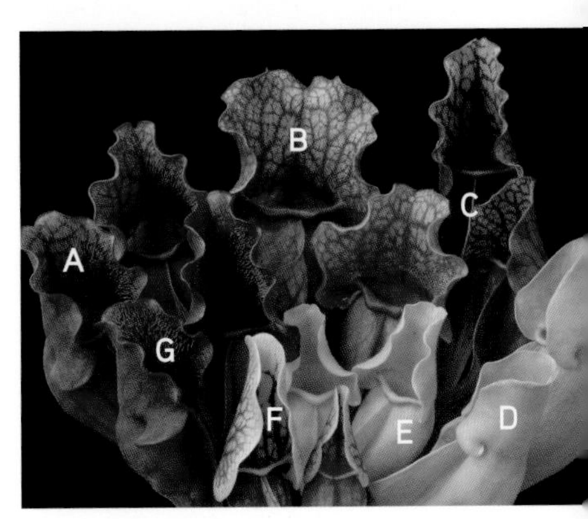

A: *S. purpurea* ssp. *venosa* v. *venosa*
B: *S. purpurea* ssp. *venosa* v. *burkii*
C: *S. purpurea* ssp. *purpurea*
D: *S. purpurea* ssp. *purpurea* f. *heterophylla*
E: *S. purpurea* ssp. *venosa* v. *burkii* f. *luteola*
F: *S. purpurea* ssp. *venosa* v. *montana*
G: *S. purpurea* ssp. *venosa* 'Red Ruffles'

Sarracenia purpurea* ssp. *venosa* var. *venosa* form *pallidiflora

This form is entirely green with yellow to white petals.

Sarracenia purpurea* ssp. *venosa* var. *burkii

Rather similar to the Atlantic coastal variety, the primary difference is the pink flowers and pale green to nearly white umbrella style. They are found on the Gulf Coast from the Florida Panhandle, through southern

Sarracenia purpurea ssp. *venosa* var. *montana*, a recently recognized variety nearly extinct in the southern Appalachian mountains.

Alabama, to southeastern Mississippi. Some botanists call this variety the species *S. rosea*.

Sarracenia purpurea* ssp. *venosa* var. *burkii* form *luteola
Same as above but the pitchers are entirely green with pale green to white petals.

Sarracenia purpurea* ssp. *venosa* var. *montana
This rare variety is found in the Appalachian mountains of the Carolinas and extreme northeastern Georgia. The short pitchers are under 5 inches (12.5 cm) in length, with a red body and an unusual collar pinched close together, rather than open and undulating, that is yellowish green with red veins. It is also very cold hardy.

Cultivars

***Sarracenia* 'Red Ruffles'**
A plant of my own selection, given to me by Don Agnostinelli of California State University in Sacramento. This plant produces short, squat, maroon leaves that are almost held upright on short petioles, and it has a highly undulating collar. 'Red Ruffles' also produces offshoots over time, producing dense clumps of pitchers.

***Sarracenia* 'Belly of Blood'**
Chosen by Barry Rice from seedlings grown from a ssp. *venosa* plant of unknown heritage. The plump pitchers have green collars strikingly veined in red; the bodies of the pitchers are deeply scarlet.

***Sarracenia* 'Phoenix'**
Discovered in the New Jersey Pine Barrens and registered by Christoph Berlanger, this red-mottled plant has a collar edged with lime green and orange-yellow flowers.

The Yellow Trumpet Plant

(SARRACENIA FLAVA)

Sarracenia flava is another highly variable pitcher plant, named for its tall flowers with pendulous, bright yellow petals—that happen to smell like male cat pee! It occurs from southern Virginia, where it is almost extinct, south along the coastal plain to extreme northern Florida, and then east to Mobile Bay, Alabama.

S. flava produces erect, tall pitchers from around 20 to 36 inches (50.5–91.5 cm) tall, and occasionally taller. The species is often one of the first to flower, sometimes as early as late February in its southernmost range. The best pitchers are usually grown in spring and summer, although some varieties continue pitcher production until early autumn. This species is one that produces secondary leaves known as phyllodia, which look similar to flat, straight iris leaves. Phyllodia usually appear in late summer and can remain on the plant through winter, long after the pitchers have deteriorated.

A devious lynx spider snatching a meal from *Sarracenia flava* var. *rugelli.*

The pitchers themselves are rather handsome. They appear as elongated, narrow funnels with a flared mouth and rather broad lip, or peristome, and a narrow neck that holds erect the large, almost horizontal lid. Unlike *S. purpurea*'s collar, this lid effectively keeps out most rain, and it acts as a landing platform for flying insects. Crawling insects follow nectar trails up the pitcher's length, particularly along the reduced ala (winglike projection) at the leaf's front seam. Various color patterns lead the prey to the most treacherous parts of the leaf. Insects appear quite intoxicated by the time they are in the vicinity of the wide mouth, under the lid, or at the neck where the foothold is rather slippery. Drunken insects fall down the narrowing tube. The beating of wings may cause a vacuum in the pitcher, sucking it down further.

The interior is so waxy smooth that insects rarely can maintain a foothold. Downward-pointing, needlelike hairs are found at its deepest point. Digestive juices are secreted by the plant in the lowest portions of the trap, and the level of this liquid rises as more insects are caught. The first victims typically drown in this fluid, which dissolves them down to their exoskeletons. Yellow trumpets may catch such enormous quantities of insects that the pitchers topple from the weight. Flies, ants, wasps, beetles, and moths are the most common prey.

Although no subspecies of *S. flava* have been named, Dr. Donald Schnell has divided the species into several naturally occurring varieties. In horticulture, many hybrids between these varieties can also be found, often muddling the distinctive qualities of this complex species.

Varieties and Forms

Sarracenia flava var. *flava*

This is the predominant variety found along the Atlantic coastal plain, especially in the Carolinas, but wetland drainage has severely reduced the once vast populations. The pitchers are green to yellow-green, with a red blotch at the throat or column that radiates into coarse veins on the underside of the lid. The exterior of the tube has fainter veining mostly along the upper parts.

The stunning varieties of *Sarracenia flava*.

Sarracenia flava var. *rugelii*

This is the most common variety found in the southern range of the species, especially along the Gulf Coast. Large and robust, the pitchers are entirely green except for a large red blotch on the column under the lid. It is often called "red blotch" or "cut throat" among hobbyists.

Sarracenia flava var. *ornata*
Highly variable and found uncommonly throughout the species' range, this striking variety is yellow-green with vivid reddish purple reticulated veining covering nearly all of the pitcher. The most densely veined forms are highly desirable in cultivation.

Sarracenia flava var. *maxima*
Found rarely throughout its natural habitat, this form is entirely green, although newly emerging leaves may be tinted pink. This variety was named in the nineteenth century but later became confused with other vigorous forms of *S. flava*.

Sarracenia flava var. *cuprea*
This beautiful variety is found naturally along the Atlantic coast. Lightly veined, the lid and upper portion of the tube are coppery red to bronze; it also has a radiating red blotch on the column. It is usually called "coppertop" in cultivation.

Sarracenia flava var. *atropurpurea*
This very rare variety of the species is found throughout its range but was once more common on the Atlantic coast. The entire exterior of the pitcher, including the lid, is red to purple. However, in cultivation the intense hue is difficult to maintain and requires full, all-day sun during the growing season.

Sarracenia flava var. *rubricorpora* in a wet savanna in the Florida Panhandle.

Sarracenia flava var. *rubricorpora*
A stunning variety known from a few populations along the Gulf Coast in Florida, its pitcher tube is entirely red, and the lid is green with red veining. A prominent dark blotch on the column may be purplish black in some extreme forms. This variety may lose its color when moved or transplanted, but the color returns. Each growing point usually produces one or two pitchers in

spring and again in late summer, separated by noncarnivorous phyllodia leaves that are also tinted red.

Sarracenia flava* var. *viridescens
This plant has been found only a few times in the wild and is entirely anthocyanin free, meaning it has no genes for any red color, even in the newly emerging leaves.

It should be noted that forms of *S. flava* from the Florida Panhandle, when self-pollinated and grown from seed, do not entirely reproduce the mother plant. This may be a sign of regressed genes at play. If you self-pollinate the variety *rubricorpora*, for example, the offspring will mature into a mixture of var. *ornata*, var. *rugelli*, and var. *rubricorpora* as well as what appears to be hybrids between of all three!

Cultivars

Legendary British nurseryman and author Adrian Slack introduced three cultivars of this species worth noting.

***Sarracenia flava* 'Claret'**
Slack describes this as being tinted maroon with heavier red veins.

***Sarracenia flava* 'Burgundy'**
A similar clone to 'Claret', it has dark red tubes with a greenish lid veined in red.

***Sarracenia flava* 'Marston Dwarf'**
Produces clumps of heavily veined pitchers not more than 12 inches (30.5 cm) high.

The Sweet Trumpet

(SARRACENIA RUBRA)

Sarracenia rubra is a species with a long and controversial history that is still debated today. Currently the plants are divided into five subspecies. Some botanists argue that a few of these are species in themselves, while others point out that the similar flowers of all tend to indicate one rather variable group. There are also plants that are clearly *S. rubra* but don't seem to fall under any yet defined group.

As a whole, the plants are generally clump-producing trumpets that are comparatively smaller in stature than some of their cousins. *S. rubras* are notable for their small, bright red to dark red flowers—some of which are fragrant to variable degrees, hence the common name. The scent of some is reminiscent of roses or, as some collectors have noted, cherry-flavored Kool-Aid. Since the rhizomes commonly branch into multiple growing points, not only are masses of pitcher leaves produced, but also the multiple flowers in spring, enhanced by their perfume, can make a showy spectacle.

All of the subspecies produce two types of pitchers. In the spring the pitchers are generally small, somewhat floppy, and snakelike, looking remarkably similar across the different forms. The summer pitchers are much more erect and robust, and all are highly veined in fine red venation.

The insect-catching mechanism is rather similar to that of the other upright trumpet species. The prey are lured by color and nectar to the area of the mouth and then fall down the lanky, narrow tube.

Subspecies

The following are general descriptions of the subspecies, but be mindful that there can be variations in some forms of the plants, mostly in the coloration of the pitchers.

Sarracenia rubra ssp. *rubra*

This most popular subspecies has highly fragrant flowers. The summer pitchers are narrow, highly veined, with a short pointed lid, and reach a height of 12 to 18 inches (30.5–45.5 cm). From eastern North Carolina and South Carolina.

Sarracenia rubra ssp. *jonesii*

This very endangered species is virtually extinct in its mountainous habitat of North and South Carolina. It is notable for its handsome pitchers, which are rather similar to ssp. *rubra* but with a noticeable bulge in their upper parts.

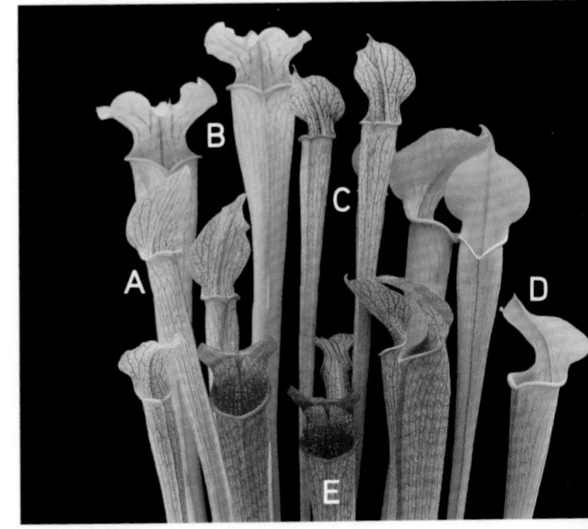

A: *S. rubra* ssp. *rubra*
B: *S. rubra* ssp. *alabamensis*
C: *S. rubra* ssp. *jonesii*
D: *S. rubra* ssp. *wherryi*
E: *S. rubra* ssp. *gulfensis*

The flowers are particularly sweet smelling and a very bright red. The pitchers can be over 24 inches (60 cm) tall. This variety is especially cold hardy. There is an all-green form called f. *viridescens* that is anthocyanin free, as well as "red" forms of this subspecies. This plant is also known as *S. jonesii*.

Sarracenia rubra* ssp. *gulfensis

A plant from the Florida Panhandle, the pitchers are tall and superficially resemble ssp. *jonesii*, but can be more variable. The large lid undulates slightly. The flowers have a weak scent. This subspecies can have variable pitchers up to 2 feet (0.6 m) tall, colored green with red veins to nearly entirely reddish purple. The form *luteoviridis* is anthocyanin free and therefore entirely green with yellow flowers.

Sarracenia rubra* ssp. *alabamensis

Also known as *S. alabamensis*, this plant is endangered and only found in a small area north of Montgomery, Alabama. The pitchers are robust and stocky, with a large lid that has very wavy margins. Up to 20 inches (50.5 cm) tall. They can also be variable in reddish coloration.

Sarracenia rubra* ssp. *wherryi

Rather similar in appearance to ssp. *alabamensis*, this plant is limited to southern Alabama. The pitchers can be 12 to 16 inches (30.5–40.5 cm) in height and variably veined and colored. The flowers are very fragrant.

The Pale Trumpet

(SARRACENIA ALATA)

Sarracenia alata occurs in two widely separated regions of the Gulf Coast. Populations occur in southwestern Alabama, southern Mississippi, and southeastern Louisiana. Farther west, it grows in western Louisiana and scattered parts of eastern Texas. Its unique flowers are the palest yellow to almost pure white, hence its common name.

The upright trumpets are quite variable in both size and coloration and can be up to 2 to 3 feet (0.6–0.9 m) in height. The pointed lids are not flared as in *S. flava* but are usually held closer to the mouth, with a wider column, showing its probable kinship with *S. rubra*.

Varieties and Forms

Largely ignored taxonomically until recently, *Sarracenia alata* has several varieties and forms that are now generally accepted.

Sarracenia alata* var. *alata

This is the most common variety throughout its natural range. The yellow-green pitchers have slight red veining on the interior of the lid and the upper portion of the pitcher tube. The exterior can sometimes have a slight fuzziness; these forms are called "pubescent." Also, vague translucent areoles may occur along the upper section of the trumpets.

Some varieties of *Sarracenia alata*. While many of these forms were seen in the wild and grown in cultivation, most did not receive names or were not described formally until Schnell and McPherson's monograph *Sarraceniacea of North America* was published in 2011.

Sarracenia alata* var. *atrorubra

Very rare, the pitchers are a vibrant red that deepens as the pitchers age. Scattered populations have been found mostly in Mississippi.

Sarracenia alata* var. *cuprea

Similar to var. *alata* but with a coppery colored lid, and very uncommon in Mississippi and eastern Louisiana.

Sarracenia alata* var. *nigropurpurea

Another rare variety with blackish purple coloration in the upper pitcher, especially along the interior of the mouth and under the lid.

Sarracenia alata* var. *ornata

The yellow-green pitchers of this variety are heavily laced with reticulated crimson veins, with fewer on the lid.

Sarracenia alata* var. *rubrioperculata

Similar to var. *ornata* but with deep red coloring in the interior mouth and the underside of the lid.

Sarracenia alata f. *viridescens*
This form is anthocyanin free and thus entirely green. Extremely rare, the first specimen was discovered by enthusiasts Carl and Melissa Mazur in Alabama in 2001.

The White Trumpet

(SARRACENIA LEUCOPHYLLA)

Considered by many to be the most beautiful of the American pitcher plants, *S. leucophylla* grows from southwestern Georgia to southern Mississippi, being most abundant in the Florida Panhandle and southernmost Alabama.

The late-winter to early-spring flowers are large, red petaled, and as tall as 3 feet (0.9 m) high. The first pitchers of the season are usually thinner and less robust than those produced later, and in summer several noncarnivorous phyllodia leaves are usually produced. It is the second crop of pitchers, grown in late summer to early autumn, that show this species in all its glory—they can be as tall as the flowers were in spring and put on a rather spectacular show at least until the first frosts of winter. The underside of the undulating lids is lined with bristly hairs similar to those of *S. purpurea*.

All varieties of *S. leucophylla* are extremely attractive to nectar-feeding insects, and the pitchers can fill nearly to the brim faster than any other species. The white coloration of some carnivorous plants may have evolved to enhance the trapping of nocturnal insects, particularly during phases of the full moon (as discussed in my article "By the Light of the Silvery Moon," *Carnivorous Plant Newsletter* vol. 39, no. 1). That *S. leucophylla* produces its most robust white leaves late in the season, coinciding with the brilliant harvest moon, underscores this idea.

Varieties and Forms

Sarracenia leucophylla **var.** ***leucophylla***
Nearly all of the plants fall under this variety but can be extremely variable. The lower pitcher tubes can be green, tan, or a purplish red. The upper portion of the tube, mouth, column, and lid are nearly pure white laced with a network of veins. In some populations the veining can be coarse;

in others they appear as a thin netting. This veining can also range from green to pink to red to purple, depending on the plant's genetic makeup.

Sarracenia leucophylla var. _alba_
This variety is similar to the preceding with all its variations except that the area of the mouth and column can be nearly pure white with little or no veining.

Sarracenia leucophylla f. _viridescens_
This form is extremely rare and is anthocyanin free, having no red coloration. Even the flowers are greenish yellow.

Sarracenia leucophylla, showing several variations in color and form.

Cultivars

***Sarracenia* 'Hurricane Creek White'**
A popular clone of var. *alba* from a once large population of the species that has since been destroyed.

***Sarracenia* 'Schnell's Ghost'**
A cultivar that appears to be an introgressed hybrid that has backcrossed with the species. It has yellow flowers and green-and-white pitchers with a lid column wider than typically seen in the species. Newly emerging leaves can also be tinted pink.

***Sarracenia* 'Tarnok'**
A clone of a var. *leucophylla* plant with curiously mutated flowers that have multiple sepals and petals (called "tepals")

Sarracenia 'Hurricane Creek White', a cultivar of the variety *alba*, now extinct in the wild.

that persist through the growing season and cannot produce seed. The pitchers are red and white. It was named for the landowner, who is preserving a large stand of *S. leucophylla* on his property, and it was introduced to cultivation by the Atlanta Botanical Garden.

***Sarracenia* 'Bris'**

A mutated clone I found among *S. leucophylla* seed-grown plants at California Carnivores; the name, suggested by Barry Rice, is taken from the Jewish ritual of circumcision. The lid is a small, abbreviated appendage overhanging the mouth.

The Mountain Trumpet

(SARRACENIA OREOPHILA)

On the verge of extinction, *Sarracenia oreophila* manages to survive in only a handful of populations in northeastern Alabama, and in the mountainous areas where Georgia and North Carolina meet. It was once more populous but always considered rare, and it may soon disappear from the wild. Small populations once existed from central Alabama to central Georgia, but these have all been extirpated. A single plant was found in north central Tennessee in the early twentieth century but was accidentally killed at a university.

The plant is unusual in that the pitchers are produced primarily in spring and early summer. By midsummer its natural habitats often become somewhat drier, and the pitchers wither and are replaced by strongly curved phyllodia. This carries over in cultivation even when the plants are kept very wet.

The flowers are yellow. The pitchers, vaguely similar to those of *S. flava* except for the more broadly opened mouth, wider neck, and somewhat dome-shaped lid, grow to about 12 to 24 inches (30.5–60 cm) tall.

Two points are notable in this rare species. One is its cold hardiness, as its mountain habitat will see snowfall more regularly than will other *Sarracenia* on the warmer coastal plain. Temperatures briefly down to 0°F (-18°C) are not unheard of. The second is the plant's capacity to add vigor to offspring when hybridized with other species of *Sarracenia*. Hybrids with *S. oreophila* can attain stunning coloration and spectacular size.

Varieties and Forms

Sarracenia oreophila* var. *oreophila

This is the most common variety found in the remaining wild populations, and the yellow-green pitchers have light red veining on the interior of the pitcher tube, column, and lid.

Sarracenia oreophila* var. *ornata

Once more common but now nearly extinct in the wild, the pitchers are strongly covered with reticulated scarlet to purple veining on both the interior and the exterior of the entire leaf. Red to purple coloring can suffuse much of the pitcher, and the mouth and column can sometimes be dark purple. In North America the strains are sometimes called "Sand Mountain" forms after the region of northeast Alabama where they originated.

Sarracenia oreophila 'Don Schnell', a cultivar of var. *ornata* chosen by the author.

Cultivars

***Sarracenia* 'Don Schnell'**

Of my own selection, this clone of an *ornata* variety is heavily veined, with a crimson nectar roll and purplish red throat.

The Hooded Pitcher Plant

(SARRACENIA MINOR)

Sarracenia minor is a curious-looking pitcher plant with a wide range, along the coastal plain from southeastern North Carolina south into much of the Florida Panhandle and all of the northern half of Florida. Recently, populations have been found as far south as Okeechobee County, making it the most southern-growing member of its genus.

The hooded pitcher plant.

The unusual pitchers of *S. minor* have an almost grinning, monkish appearance. In most of its range the leaves average around 12 inches (30.5 cm) in height. The tubes are smooth with a wide ala at the front seam. The lid of the pitcher forms a domed canopy over the mouth. On the upper portion of the back side of the pitcher are many opaque light windows. The pitchers are generally green, with a coppery red coloration along the upper parts when grown in full sun.

This pitcher's trapping of prey is rather unique. When insects are led to the lip of the mouth by nectar trails, they find themselves in a rather darkened position, due to the overhanging hood. Crawling insects are encouraged to enter the trap, where it is much brighter, due to sunlight shining through the windows. Flying insects are fooled into believing the windows are escape hatches. The crawling prey find no foothold on the waxy interior of the hood, while the flying ones slam into the windows for a stunned surprise. Either will tumble helplessly down the narrowing tube and into digestive juices below. In the wild, hooded pitcher plants seem most attractive to ants, although a large variety of flying insects are lured and eaten, too.

The drooling nectar of *Sarracenia minor* lures insects to the edge of doom, while the light windows offer a false hope of escape.

The flowers are medium-size and a pleasant, buttery yellow. Unique among *Sarracenia*, the spring flowers often open simultaneously with the first leaves of the season.

Varieties and Forms

Sarracenia minor* var. *minor

This is the predominant variety found throughout its range, with pitchers up to 12 inches (30.5 cm) tall, although they can be shorter or taller. There is some variability among the plants; some populations have thinner, stiffer pitchers, whereas others may be plump and soft. Most will become coppery red in the upper parts when grown in full sun.

Sarracenia minor* var. *minor* f. *viridescens

This form is very rare and, being anthocyanin free, lacks any red coloration.

Sarracenia minor* var. *okefenokeensis

This variety is named after southern Georgia's Okefenokee National Wildlife Refuge where most are found. Its primary feature is the enormously tall and narrow pitchers that can reach up to 3 feet (0.9 m) high in cultivation; even taller specimens have been found in the wild. It often grows on floating hummocks of live sphagnum moss and similar very wet substrates. *S.* 'Okee Giant' is a name I registered before Dr. Don Schnell formally described the variety *okefenokeensis*. Both terms are valid.

One problem with the hooded pitcher plant, in my experience, has been the lack of seed production when the flowers are self-pollinated. To get seed set, it is best to cross-pollinate the flowers with different genetic clones. This is one reason why seed-grown plants of the variety *okefenokeensis* are so rare—few clones exist in cultivation. Hybrids between the two varieties are often encountered and should be labeled so.

The Parrot Pitcher Plant

(SARRACENIA PSITTACINA)

Unlike any other plant in its genus, *Sarracenia psittacina* seems to have more in common with its distant Pacific coast relative *Darlingtonia* than with a typical trumpet plant.

Parrot pitcher plants have an affinity for wetter, low-lying areas of swampy savannas, and they are often flooded by heavy rains. The species grows along the coastal plain from extreme southeast South Carolina through scattered areas of southern Georgia, and much of the Florida

Panhandle, through southernmost Alabama, Mississippi, and extreme southeastern Louisiana.

The pitcher leaves are decumbent, lying in a rosetted pattern pressed along the ground or slightly raised into the air. The small flowers are red and have a mild, sweet aroma. The elongated tubes have a large, wavy ala and end in a hollow, puffed hood that is rather beaked—hence its common name. Under the beak, where the nectar-baited ala ends, is a small circular opening. Inside the inflated hood, this opening is surrounded by a collar, making it not unlike a minnow trap or lobster pot. The back side of the hood and upper part of the narrow tube are laced with numerous light windows, similar to those of *S. minor* and *Darlingtonia*. The interior of the tube is lined with extra-long, intermeshed, needlelike hairs, all pointing toward the base of the leaf.

Prey caught by this plant suffer a hideous death. Once they are inside the hood, the exit is difficult to find due to the puckered collar. Insects thus enter the brighter tube lighted by the windows. However, there is no possible retreat, for to back out means to be painfully pierced by the numerous needlelike hairs. The victim has no choice but to proceed into the digestive acids in the lower part of the pitcher.

We know that the parrot pitcher plant catches aquatic animals when underwater thanks to the many tadpoles and other swimming creatures found in the leaves after a flood. When the water recedes, the plant resumes catching ants, slugs, and other crawling things. Flooding the plant in cultivation is not necessary for good growth.

A pitcher of *Sarracenia psittacina* sliced in half, showing the interior collar entering the beaked head and needle-lined hairs of the tube—a torturous death to those who enter!

Varieties and Forms

Sarracenia psittacina* var. *psittacina

This is the typical parrot pitcher plant found throughout its range, with little variation. The leaves average around 6 to 8 inches (15–20 cm) long. The coloration of the plant can be mostly green when shaded by low-growing vegetation or with much red venation when grown in full sun.

The parrot pitcher plant *Sarracenia psittacina*.

Sarracenia psittacina* var. *psittacina* f. *viridescens

The all-green form of the variety with greenish yellow flowers, being anthocyanin free.

Sarracenia psittacina* var. *okefenokeensis

This impressive variety is named after the Okefenokee Swamp, where it is most abundant, often growing on waterlogged mats of sphagnum moss with its cousin *S. minor* 'Okee Giant'. Less commonly, it is also found in adjacent areas of Georgia and in the Florida Panhandle and southern Alabama. The leaves can reach 12 inches (30.5 cm) in length and have globose, golf ball–size hoods.

Sarracenia psittacina* var. *okefenokeensis* f. *luteoviridis

Identical to the above but anthocyanin free and thus green with yellow-green flowers.

Hybrids

American pitcher plants are unusual in the plant world because the species can be readily hybridized, and these offspring are not sterile, as with most other plants, but are capable of being self-pollinated or hybridized even further.

This situation can be vexing to the field botanist, as it was in the last century, when many natural hybrids were thought to be species and in fact were given Latin names. But to the horticulturist, the ease of hybridizing *Sarracenia* is both fun and exciting, as the results are often beautiful. Further, when one takes into account the already intriguing forms of some of the more complex crosses already existing and projects into the future the additional possibilities as hybridization programs become more serious and popular, the results will probably be utterly fantastic. As with orchids and roses and African violets, there is no end to the possibilities. Carnivorous vegetable gargoyles may be the future of *Sarracenia*, with carefully selected breeding.

In the wild, where two or more species are found growing together, the differing flowering times usually keep hybrids in check. But now and then the flowers of different species coincide, and the result may be scattered hybrid offspring or the occasional hybrid swarm. Add to this future backcrossing, and the result in some stands may be pitcher plants of very confused ancestry.

Simple hybrids are crosses between two species, many of which are found in the wild as well as created in horticulture.

Because some of the common natural hybrids were once given Latin names, as if they were a separate species, these names have remained in use. These are names of convenience, and most modern growers will also know them by their correct species name. Thus *S. leucophylla* × *rubra* is also called *S.* × *readii*. (See the Simple *Sarracenia* Hybrids sidebar on page 113.)

I would strongly suggest that anyone hybridizing the plants be careful to keep good labels and records of the crosses. Eventually, pedigree will not be so important to the collector who simply admires a beautiful plant. But even if a plant is merely coded or numbered on its label, do keep records of as much known data about the history of the plant as possible. For helpful tips on recordkeeping, see page 114.

Actual pollination techniques will be covered under "Propagation." Here I will discuss what you can expect with the general results of hybridizing, as well as a few simple descriptions of some of the more popular representative crosses.

Remember that under the laws of genetics, 50 percent of the offspring will generally look intermediate between both parents, while the remaining plants will lean more toward one parent or the other. Variability is the rule, and extremes are certainly common, particularly when the ancestry of a plant is complex.

SIMPLE *SARRACENIA* HYBRIDS

Following are the Latin names of simple *Sarracenia* hybrids found in the wild. Although these names are considered outdated, they are still frequently encountered in horticulture.

S. x *catesbaei*	=	*S. purpurea* x *flava*
S. x *moorei*	=	*S. flava* x *leucophylla*
S. x *popei*	=	*S. flava* x *rubra*
S. x *harperi*	=	*S. flava* x *minor*
S. x *mitchelliana*	=	*S. purpurea* x *leucophylla*
S. x *exornata*	=	*S. purpurea* x *alata*
S. x *chelsonii*	=	*S. purpurea* x *rubra*
S. x *swaniana*	=	*S. purpurea* x *minor*
S. x *courtii*	=	*S. purpurea* x *psittacina*
S. x *areolata*	=	*S. leucophylla* x *alata*
S. x *readii*	=	*S. leucophylla* x *rubra*
S. x *excellens*	=	*S. leucophylla* x *minor*
S. x *wrigleyana*	=	*S. leucophylla* x *psittacina*
S. x *ahlesii*	=	*S. alata* x *rubra*
S. x *rehderi*	=	*S. rubra* x *minor*
S. x *gilpini*	=	*S. rubra* x *psittacina*
S. x *formosa*	=	*S. minor* x *psittacina*

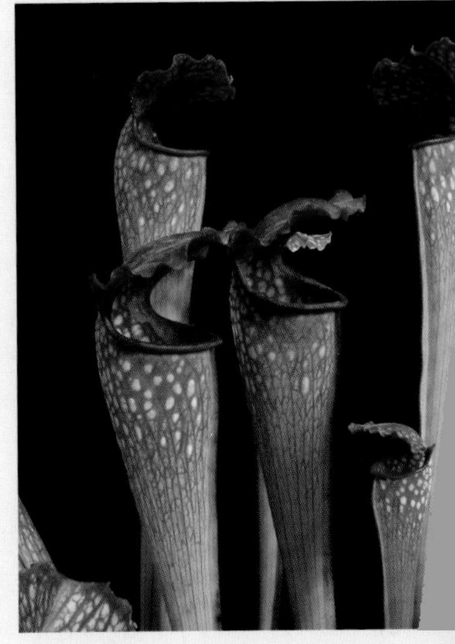

Sarracenia x *excellens*, the hybrid between *S. leucophylla* and *S. minor.*

RECORDKEEPING

Following are suggested codes for use with *Sarracenia*. I have slightly amended Adrian Slack's codes from his book *Carnivorous Plants*. The general idea is to reduce the Latin name to one or two letters. Where subspecies, forms, and varieties are involved, an additional letter or two is added. Space does not allow me to list all of the varieties.

a	=	*alata*
an	=	*alata* var. *nigropurpurea*
f	=	*flava*
fa	=	*flava* var. *atropurpurea*
fc	=	*flava* var. *cuprea*
ff	=	*flava* var. *flava*
fo	=	*flava* var. *ornata*
fr	=	*flava* var. *rugelli*
fru	=	*flava* var. *rubricorpora*
l or ll	=	*leucophylla* var. *leucophylla*
la	=	*leucophylla* var. *alba*
lr	=	*leucophylla* 'Red and White'
o	=	*oreophila*
oo	=	*oreophila* var. *ornata*
osm	=	*oreophila* 'Sand Mountain'
m	=	*minor*
mog	=	*minor* var. *okefenokeensis* 'Okee Giant'
p	=	*purpurea*
pp	=	*purpurea* ssp. *purpurea*
ps	=	*psittacina*
pso	=	*psittacina* var. *okefenokeensis*
pv	=	*purpurea* ssp. *venosa*
pvb	=	*purpurea* ssp. *venosa* var. *burkei*
r	=	*rubra*
ra	=	*rubra* ssp. *alabamensis*
rg	=	*rubra* ssp. *gulfensis*
rj	=	*rubra* ssp. *jonesii*
rr	=	*rubra* ssp. *rubra*
rw	=	*rubra* ssp. *wherryi*

The Simple Hybrids

Many of these can be found in the wild. All are greatly influenced by the subspecies, forms, or varieties of the parents. The following are plants with rather unique characteristics.

Sarracenia purpurea* × *flava
This is probably the most popular hybrid, although it is not a particularly good insect catcher, and the pitchers can topple if they collect rainwater. The plump pitchers curve upright, the undulating hood highly influenced by its *S. purpurea* parent. The flowers are usually pale red.

Sarracenia purpurea* × *leucophylla
Similar to × *flava*, but much more colorful in the hood, which can be ruffled and mottled with pinks, whites, and reds, and highly veined. Red flowers.

Sarracenia purpurea* × *minor
An often reddish plant with low-growing curved pitchers that all face inward, and a monklike hood overhanging the mouth. The flowers are a dark orange-red.

Sarracenia purpurea* × *psittacina
Strange, dark red, ground-hugging pitchers, with an unusual hood that curls inward on either side.

Sarracenia flava* × *leucophylla
Very handsome trumpets, mottled with extra color in the hood, and with orange flowers. Excellent pitchers throughout the season.

Sarracenia leucophylla* × *rubra
Clumps of narrow, colorful pitchers with ruffled lids, heavily influenced by the subspecies of *S. rubra* involved. Many red flowers.

Sarracenia leucophylla* × *minor
A very popular cross, with an undulating, overhanging lid, heavily dappled in reds and whites. Orange flowers.

Sarracenia leucophylla × *psittacina*
Extremely bizarre, the pitchers are curved with beaked heads. A poor insect catcher, but often very bright and colorful.

Sarracenia rubra × *flava*
These crosses usually produce handsome trumpets with rich red venation, and smallish, pale-red flowers.

Here are some general guidelines concerning the simple hybrids.

Most crosses between the upright trumpet species (*S. flava, S. leucophylla, S. rubra, S. alata*, and *S. oreophila*) produce attractive, tall, and colorful pitchers.

Hybrids with *S. psittacina* can be very weird and low-growing, and are often rather unattractive, as well as poor insect catchers.

The trumpet varieties crossed with *S. minor* inherit the overhanging hood of that species and thus can be rather handsome. Light windows can also be inherited.

For color, nothing beats the influence of *S. leucophylla*.

S. oreophila crosses are almost always extra vigorous and beautiful.

Keep in mind the seasonality of the parents. *S. flava, S. oreophila*, and *S. minor*, for example, produce their best pitchers early in the year. *S. leucophylla, S. rubra*, and *S. alata* are often best in later summer. Crosses between these two groups result in strong pitchers throughout the season. *S. purpurea* and *S. psittacina* hold their leaves through winter in fairly good shape and can influence offspring.

When a hybrid is labeled F-2, this means it resulted from a hybrid parent that was self-pollinated, and extremes in variability can be expected.

The more cold-hardy species can produce cold-hardy hybrids: *S. purpurea* ssp. *purpurea, S. rubra* ssp. *jonesii*, and *S. oreophila* pass on cold tolerance to their offspring.

Although the rules have recently changed, the plant producing the seed is usually listed first in hybrid equations. New rules allow them to be alphabetized.

Complex Crosses and Greges

In the appendix, under the Taxonomy section (page 354), I discuss the problem faced by carnivorous plant growers on the use of grex names for complex hybrids. I will only suggest that when labeling or recording plants that you have crossbred yourself, try to use brackets and parentheses, and use codes as the formula lengthens.

Here is an example of the progress of such hybridization.
First cross: S. *purpurea* × *flava*.
Second cross: S. (*purpurea* × *flava*) × *leucophylla*.
Third cross: S. [(*p* × *f*) × *l*] × *minor*.

Cultivars

Cultivar means "cultivated variety."

A cultivar can be any outstanding individual plant species or hybrid that is so desirable in its attributes that it should be forever propagated only vegetatively—through division or tissue culture—to preserve its genetic makeup and guarantee it is an exact duplicate or clone of the originally selected plant. One should never propagate a cultivar through self-pollinated seed, as most of the offspring will be variable and will not duplicate the mother plant. If one does so, the cultivar name should never be put on the seedlings—use the species formula or grex name and indicate them as F-2 (or self-pollinated) offspring.

In late 1998, the International Society of Horticultural Science (ISHS), which oversees the International Cultivar Registration Authority (ICRA), appointed the International Carnivorous Plant Society (ICPS) as the official registrar of carnivorous plant cultivars.

Unfortunately, according to their rules, the ICRA actually allows cultivars to be reproduced through self-pollination if the offspring appears like the mother cultivar, but all breeders I know insist that their cultivars be duplicated only vegetatively.

Cultivars should be registered with the ICPS in order to be considered legitimate. You will need to document the plant with a photo, description, and history, if known. The plant can also be described in such a way in a book or a catalog, but it will not be considered legitimate until registered with the ICPS. Cultivars should be given a fancy name, but not in Latin. The name should be preceded by the letters cv to identify it as a cultivar, or the name should be put into single quotes. The ICPS, on their website www.carnivorousplants.org, has excellent information on how to register a cultivar, including the forms to fill out and submit.

Sarracenia 'Abandoned Hope'.

Due to the explosion of popularity of the CP hobby in recent years, there has been a proliferation of registered cultivars from growers around the world. These are often published in the *Carnivorous Plant Newsletter*, and all cultivars are listed with photographs on the ICPS website.

Unfortunately, many *Sarracenia* cultivars remain rare or very expensive when available, because cloning American pitcher plants via tissue culture is a difficult challenge taken on by very few. There have been a few great successes with plants achieving wide distribution, while rarer cultivars can only be propagated the old-fashioned way, through division.

Hybrid Cultivars

I have already mentioned cultivars of the species earlier in this chapter. Here we will discuss cultivars of hybrid origin.

Sarracenia × 'Willissi'

Of very confused ancestry, this plant was produced by Veitch and Sons nursery in England in the nineteenth century. A beauty, it has curved, upright pitchers with a ruffled lid similar to a *S. purpurea* × *leucophylla* cross, but the pitchers are pink initially, turning a deep plum red. Probably the only surviving *Sarracenia* cultivar from the nineteenth century in wide circulation.

Sarracenia 'Leah Wilkerson'

Named by Brooks Garcia after the owner of the property where this exceptional clone was found. A natural hybrid of *S. flava* var. *rugelli* × *leucophylla*, this plant is very popular for its vibrant color and clump-forming vigor.

Sarracenia 'Royal Ruby'

A tall, stunning natural hybrid of *S. leucophylla* × *flava* that is entirely burgundy red with some pale areolas along its upper parts. Introduced to cultivation by Phil Faulisi.

Sarracenia 'Royal Ruby'.

Sarracenia 'Lovebug'

Larry Mellichamp, of the University of North Carolina at Charlotte, has produced many cultivars over the years. Among the most popular, this is a cross of [ps × p) × m] × [(p × l) × rj]. The curved red pitchers with areoles have a domed canopy lid with wavy margins.

Sarracenia 'Adrian Slack'

Bob Hanrahan operated World Insectivorous Plants in the 1970s and 1980s and introduced many rare CPs into cultivation during that time. While visiting his preserve of *Sarracenia* in southern Alabama in 1997, I was so impressed with an unknown hybrid he grew there that he honored me with the task of removing a division of the plant, which Bob submitted to tissue culture. Alas, all attempts to clone the plant failed in flask, but over the years divided plants have been circulated. When the *CPN* dedicated an

Sarracenia 'Adrian Slack'.

issue to British horticulturalist Adrian Slack, I nominated Bob's hybrid to be named *Sarracenia* 'Adrian Slack'. A rainbow of colors, the pitchers turn from green to variegated reds in its upper tube with a blood red lip and an arresting white-topped lid streaked with red veins. Stunning.

The following four cultivars were produced by Adrian Slack, who owned Marston Exotics nursery in England before an unfortunate illness caused his early retirement.

Sarracenia 'Evendine'
A clone of Slack's cross of *S. leucophylla* × (*flava* × *purpurea*). The pitchers at first are golden green and veined, later turning dark red.

Sarracenia 'Judy'
This is *S. minor* × (× *excellens*), with a highly domed lid and many light windows.

A praying mantis eating a bug on *Sarracenia* 'Lynda Butt'.

Sarracenia 'Lynda Butt'
Of unpublished ancestry. The pitchers are tall and narrow, with pale areoles around the upper tube, and a short, red-mottled lid.

Sarracenia 'Marston Clone'
A vigorous cultivar of the popular hybrid *S.* × *moorei*, with pale orange flowers.

Of my own cultivars chosen at California Carnivores, the following are noteworthy.

Sarracenia 'Abandoned Hope'
This monstrous plant is a clone of a particularly vigorous *S. purpurea* ssp. *venosa* var. *burkei* × *flava*. It is highly veined and turns very crimson as the pitchers age.

Sarracenia 'Judith Hindle'

One of the most beautiful cultivars yet produced, this plant was bred and grown by Alan Hindle, in England, and given to me as a gift; I named the plant after his wife. The compact pitchers have a wildly undulating hood; they start out green, dappled in whites and yellows. As they age, they transform to deep, dark red. The parents of *S.* 'Judith Hindle' were two separate clones of *S.* (p × f) × l.

If you wish to grow only one American pitcher plant, the vigorous cultivar *S.* 'Judith Hindle' should be your choice!

Sarracenia 'Lamentations'

Of unknown origin, this hybrid has narrow, deeply maroon trumpets with upright, pointy lids.

Sarracenia 'Extreme Unction'

Chosen from a cross of *S. minor* × [(p × f) × f], the plump pitchers are green, with a large domed lid netted scarlet, with cathedrallike windows along its back. The name is taken from the death rites of the Roman Catholic church.

Sarracenia 'Godzuki'

Made by Leo Song when he worked at California State University; I circulated it widely as *S. oreophila* × *minor* before I named and registered it after the son of Godzilla. A compact, clumping, bronzy red plant, notably lizardlike in the mouth and lid. The yellow flowers can last for over three weeks.

Sarracenia 'Godzuki', named after the son of Godzilla.

Sarracenia 'Leo Song', a cultivar of *S. oreophila* x *purpurea* ssp. *venosa*, named by the author for Mr. Song, who created it.

Sarracenia 'Leo Song'

I named this after the longtime coeditor of *CPN*, who hybridized *S. oreophila* × *purpurea v.* This colorful, clumping clone has curved pitchers with an enormous collar suggesting a screaming frilled lizard.

Sarracenia 'Deep Throat'

In the early 1990s I crossed *S. leucophylla* 'Red and White' with *S. minor* var. *okefenokeensis*. Nearly all the progeny were large vigorous forms of *S.* × *excellens*, with a few exceptions: this clone is one of them. A hulking plant with a gaping mouth and curvaceous red lips, it becomes highly variegated and brick red in its upper parts. (Pictured on the front cover.)

John Hummer, a grower from Virginia, has registered a number of cultivars he has produced. Here are three:

Sarracenia 'Hummer's Hammerhead'

This unusual plant is a clone of *S.* (*psittacina* × *rubra* ssp. *alabamensis*) × *rubra* ssp. *alabamensis*. Compact, with leaves under 8 inches (20 cm) long, the upper tube has many light windows, but its most prominent feature is the very elongated, triangular green lid.

Sarracenia 'Frogman'.

Sarracenia 'Frogman'

A cross of *S. rubra* ssp. *alabamensis* × *S. minor* var. *okefenokeensis*.

Sarracenia 'Super Green Giant'

John crossed the anthocyanin-free forms of *S.* (*psittacina* × *rubra* ssp. *jonesii*) × *rubra* ssp. *gulfensis* and created this all-green plant that contrasts attractively when grown among other more red-colored pitcher plants.

Matthew Soper, owner of the multi-award-winning Hampshire Carnivorous Plants in England, has produced two cultivars worth noting:

Sarracenia 'Juthatip Soper'
Named for Matthew's wife, this cultivar is *S.* × *mitchelliana* backcrossed with *S. leucophylla*. The upper pitcher is fuchsia red with darker veins and a wildly undulating, ruffled lid.

Sarracenia 'Johnny Marr'
Named after the English guitarist, this is an award-winning hybrid of (*S. flava* var. *cuprea* × *purpurea* var. *venosa*) × *flava* var. *cuprea*. The coppery colored pitchers turn dark burgundy with nearly black veins as they age.

Phil Sheridan is the director of the nonprofit Meadowview Biological Research Station in southern Virginia, where they work to preserve the last remnants of wild *Sarracenia* in that state. Among the many species and hybrids they grow and sell, the following two contrast nicely.

Sarracenia 'White Knight'
An offspring of *S. alata* × *leucophylla*, this is most unusual for its white flowers and white mouth and lid lightly penciled in pale green veins.

Sarracenia 'Red Viper'
Of unknown heritage, *S.* 'Red Viper' is a cultivar produced by Meadowview Biological Research Station, dedicated to preserving our vanishing populations of *Sarracenia*.

Sarracenia 'Red Viper'.

CULTIVATION

(See Parts One and Two for further details.)

SOIL RECIPE: *Sarracenia* thrive in a mix predominantly made of sphagnum peat moss. They do very well in a ratio of 70 percent peat to 30 percent horticultural sand. Another great mix is 80 percent peat to 20 percent perlite. Since perlite is slightly alkaline, I would avoid a higher percentage in the soil. Long-fibered sphagnum is also great but can be rather expensive.

CONTAINERS: Plastic pots or glazed ceramics are best. They may be drained or undrained. Young plants do well in 4-inch (10 cm) pots, mature plants in 6- to 8-inch (15–20 cm) pots or larger.

WATERING: Use the tray method for drained pots. Keep the soil permanently damp to very wet.

LIGHT: Full sun to mostly sunny is best.

CLIMATE: All *Sarracenia*, with one exception, are warm temperate, enjoying warm summers and chilly winters; they are tolerant of light frosts and brief freezes. *S. purpurea* ssp. *purpurea* can survive cold temperate climates and extended deep freezes but can also adapt to warm temperate areas with milder winters. Avoid having the pots overheat in hot summer areas, which can stress or kill *S. purpurea* ssp. *purpurea*.

DORMANCY: All require three to four months of winter dormancy, with reduced temperatures and photoperiod. The plants usually go dormant on their own as the photoperiod shortens in the autumn and return to growth in very late winter and spring, as daylight periods lengthen and temperatures warm up.

OUTDOORS: *Sarracenia* do well in temperate, warm temperate, and Mediterranean climates. *S. purpurea* ssp. *purpurea* does best in temperate and cold temperate climates but can thrive in warm temperate areas if the soil doesn't overheat. The southern species survive cold temperate climates in bog gardens mulched in winter.

BOG GARDENS: *Sarracenia* are among the best for bog gardens. (See the preceding Outdoors information.) Mulch in colder zones.

WINDOWSILLS: Good candidates for only the sunniest windows or solariums, but respect their winter dormancy. In order of most suitable first: *S. purpurea* ssp. *venosa, S. psittacina, S. rubra, S. minor, S. flava, S. oreophila, S. alata*. The one requiring the most sun is *S. leucophylla*. All trumpet plants can be superb housefly catchers.

TERRARIUMS: Generally poor candidates due primarily to their size and dormancy requirements. However, good seasonal candidates are *S. purpurea, S. psittacina, S. rubra* and their hybrids, or young plants. Better under high-intensity lights. Dormancy must be respected.

Sarracenia flava.

GREENHOUSES: Excellent in cold houses, cool houses, and warm houses, and cold frames in warm temperate climates. Even *S. purpurea* ssp. *purpurea* can survive dormancy without frost and freezes.

FEEDING: Outdoors they are often gluttonous pigs, devouring ants, flies, wasps, beetles, and moths. Can usually be hand-fed crickets, sow bugs, or dried insects.

FERTILIZING: Outdoor plants certainly don't need fertilizers, since they devour so many insects. However, where insects are scarce, such as in a greenhouse, a foliar fertilizer applied once or twice monthly during the growing season is a requirement for vigorous growth and health. Baby plants can achieve maturity in as short a time as three or four years with fertilization.

continued

TRANSPLANTING: Potted specimens should be divided and transplanted every three to five years, or when the growing points become crammed along the edge of the pot. This should be done only during dormancy or early spring growth. If you divide mature rhizomes, it is best to cut off any emerging flower buds so the plant can put its energy into healing and pitcher growth.

GROOMING: Older pitchers turn brown from the tops down; you can trim off the dead parts, but insect-filled leaves that are still healthy will continue to feed the plant. Usually all remaining pitchers will look pretty decomposed by late winter and can be removed before or during early spring's growth.

PESTS AND DISEASES: Primary pests of American pitcher plants are aphids, scale, thrips, and mealybug. Use an appropriate insecticide. Treatment may need to be repeated with annoying pests like mealybug, which can be spread by ants in greenhouses, and where natural predators of the pest are absent. Outdoors the major problem pest is aphids, which can deform newly developing pitchers rather quickly. After treatment, cut off any badly deformed emerging pitchers so the plant can put its energy into growing new, healthy leaves. Seeds can be attacked by damping-off fungus. Treat with a fungicide.

PROPAGATION

SEED: *Sarracenia* flowers are not only among the most beautiful in nature but also cleverly designed to enhance cross-pollination. Bees are their most common pollinator. Carrying pollen from a previous flower, the bees can enter only over the female stigmas between the petals, where the pollen is deposited. Once inside, they become dusted with fresh pollen. To exit, they push through the petals, thus avoiding the stigmas and self-pollination.

When the flowers open, they will remain in petal for seven to ten days on average. By lifting a petal, you can see the male anthers on the inside ceiling of the flower. The anthers, when ripe, will release their yellow, powdery pollen onto the floor of the upside-down umbrella style.

To self-pollinate the plants, collect this with a small brush or the tip of your pinky finger and deposit the pollen onto the five female stigmas. You will see these as five tiny bumplike hooks on the inside tips of each point of the umbrella style. Deposit a dab of pollen onto each one.

To cross-pollinate, carefully deposit the pollen onto the stigmas of another opened flower.

Be sure to label the flowers. Usually the pollinated plant that will produce the seed is listed first on the tag's label.

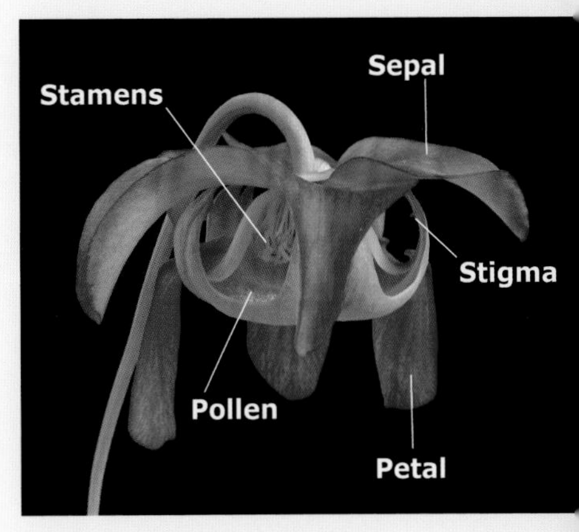

Sarracenia flower parts.

Pollen can be stored for several weeks in the refrigerator in packets of aluminum foil. It can be used on flowers that open later in the season.

When the petals drop off, the sepals and umbrella style remain, and may gradually lift upward, rather than remain nodding and upside down. The ovary, on the ceiling of the seedpod where the anthers were, will gradually swell over summer.

By autumn, the seedpod will turn brown and gradually crack open. The seed, up to several hundred, can be collected at this time. Each seed is brown to reddish tan and about the size of a large pinhead. Separate the seed from the ovary and store dry in a small paper envelope or an airtight plastic bag. Always store seed in the refrigerator.

To germinate, the seed needs several weeks of chilly, damp stratification. It is usually best to sow the seed around February, onto its preferred soil mix or milled sphagnum. Do not bury the seed. Sow sparsely, and treat with a fungicide if you see any hint of damping-off or botrytis disease. Light frost is helpful during stratification.

After stratification, with warmer, partly sunny conditions, the seed will germinate. At the end of one growing season, the seedlings will have pitchers 1 to 2 inches (2.5–5 cm) tall. The plants will take, on average, about five years to reach maturity. This can be shortened with fertilization.

continued

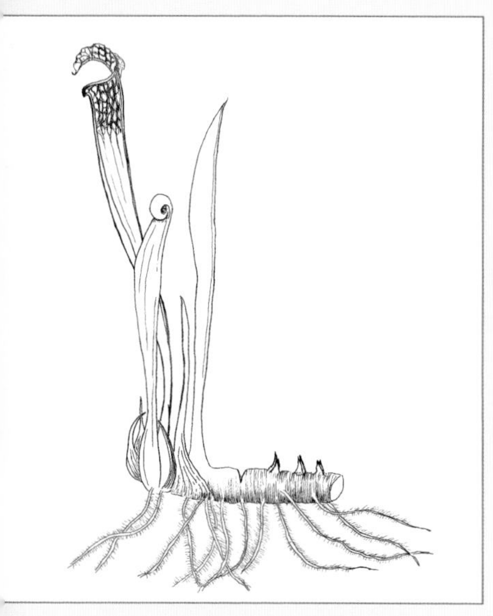

Notching a *Sarracenia* rhizome will usually encourage new growing points.

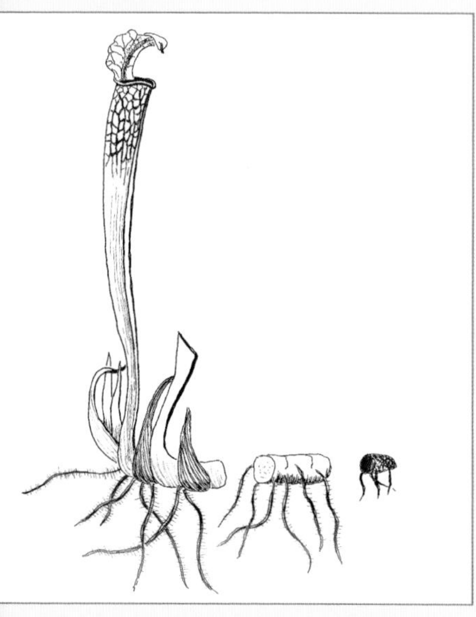

Division of a *Sarracenia* rhizome.

Propagation, *continued*

DIVISION: Most mature *Sarracenia* readily produce offshoots and new growing points year after year. Some, such as *S. purpurea* and *S. psittacina*, are slow at this or never multiply, but others, such as *S. rubra*, may develop from one growing point to more than fifty in a few short years. These clumps of plants can easily be divided.

This is best done during winter dormancy or early spring as the plants come back into growth. To divide a plant in summer or autumn will shock it, and it may not recuperate until the following year.

Remove the plant from its pot and wash away as much soil as possible. You will be able to clearly see the separate growing points from which the pitchers emerge. The rhizome is often gnarled and branching, the roots tough and wiry. By wiggling the growing points you can usually make out where they join the main rhizome. Snap these apart or cut them with a knife. Make sure the separated growing point has a few of its own roots.

If the divided plant had already begun growth, you may want to soak the divisions in a vitamin B1 solution such as SUPERthrive to overcome shock.

A healthy rhizome is white inside, similar to a russet potato. Often older parts of the rhizome are brown and dead; these should be trimmed away.

Long, branching rhizomes with few growing points may be cut further into pieces about 2 to 3 inches (5–7.5 cm) long, even if no growing points appear on that section. When potted up, these sections will produce offshoots.

New growing points can be instigated in potted *Sarracenia* when the rhizomes are long and old. Clear away some soil along the top of the rhizome to expose it. With a sharp knife, cut into the rhizome about halfway. Several cuts or notches may be made. New growing shoots will usually appear along these cuts. The following year the plant can be divided.

When repotting divisions, make sure you plant the rhizome pieces horizontally with the roots downward, the growing points at the soil surface. Cut off any emerging flowers from newly divided *Sarracenia*, as they will exhaust the plant.

TISSUE CULTURE: *Sarracenia* can be introduced in vitro through sterilized seed. Newly emerging flower buds and leaves can also succeed, but often with great difficulty. Promising new methods are being developed to enhance this process. This will greatly affect the propagation of cultivars, which have generally been in short supply.

The Cobra Plant

(Darlingtonia californica)

Light windows brighten the interior of the cobra plant's hood, inviting insects to their death.

With its bulbous green heads, twisted red tongues, and long, tubular pitchers, the cobra plant is very suitably named.

The plant was discovered by botanist J. D. Brackenridge in 1841 just south of Mt. Shasta, California. On an expedition that was exploring the northwest, Brackenridge was reported to have just collected the plant when his group had an altercation with the region's Native Americans. Fleeing the area with his arms full of serpentine leaves, Brackenridge noticed a butterfly fluttering persistently after the bobbing heads, so attractive was the plant to its insect prey. John Torrey, who described it in 1853, named the plant after his friend Dr. William Darlington.

Cobra plants are native only to Northern California and southern Oregon. Oregon populations are primarily coastal and at sea level, usually in sphagnum seep bogs or along lake margins in the heavily wooded hills. In California, vast stands of the plant can be found in Del Norte County along the Smith River basin just south of the Oregon border and further south near the coast of Humbolt County. Much further inland, cobra plants are primarily mountain plants in the higher elevations of the Trinity Alps, Mt. Shasta, and the Klamath mountains, with one population in the Sierra Nevada near Lake Tahoe. At the highest elevations, over 8,000 feet (2,438 m), the plants are covered with snow in the winter, whereas coastal stands of the plant may see frost only rarely. The climate in which they grow could be considered Mediterranean to the extent that much of the precipitation falls in winter, while summers are mostly dry. Depending on the altitude, summer days may be hot inland or cool along the coast, but for native cobras summer nights are almost always chilly.

Cobra plants may be found growing in sphagnum bogs in sunny patches of coastal forest, wet grassy meadows in the mountains, or on gravelly slopes of serpentine rock where water is constantly seeping. Typically, cobras colonize areas where springs are located and the water is usually cool and slowly moving. This is a key to understanding their cultivation. However, there are *some* inland populations where the wet soil can be rather warm.

Darlingtonia is very closely related to *Sarracenia*. There are striking similarities between it and *S. minor* and *S. psittacina*. Seedling plants are tiny, with snakelike pitchers hugging the ground like little tubes with pointy tongues hanging from their ends. When two to three years of age, the pitcher leaves tend to become more characteristic of those of mature plants.

The leaves arise from the rhizome in tubular fashion, often with a gradual twist. The head of the leaf is a puffed hood of transparent windows. Under the hood is a circular opening, difficult to distinguish from almost any angle but below. It is from the front portion of the hole that the twisted, mustachioed "tongue" emerges, also known as the "fishtail appendage."

Most of its insect prey are lured to the plant by its colorful, nectar-baited tongue. Crawling insects also follow nectar trails that run up the pitcher's exterior. The nectar is the heaviest at the base of the tongue's attachment to the circular opening. At this point, insects may be lured to enter the hole by the brilliance of the sun shining through the hood's transparent light windows. Entering is the prey's fatal mistake, for once inside the hood, escape is virtually impossible. An inner collar surrounding the entrance is similar to a lobster or minnow trap, making it almost impossible to find an exit. Instead, the victim, most often a flying insect, is attracted to the windows in the hood, perceiving them (falsely) to be skylights offering escape.

The confused insect bounces around until it tumbles helplessly down the tube. It is not yet known whether any drugs assist in the capture of prey. Regardless, hope for an escape is dismal, as the lower portion of the interior pitcher is lined with sharp, slippery, downward-pointing hairs.

Darlingtonia, unlike its cousins the *Sarracenia*, produces no enzymes to aid in the digestion of insects. But as insects become caught, the pitcher secretes water that drowns the victims. Bacteria and other microorganisms help break down the soft parts of the prey.

Native populations of cobra plants play host to two types of creatures that appear to have a symbiotic relationship with the plant. One is a tiny fly that can enter and exit the pitcher at will. Eggs are laid on the dead, trapped insects, and the small wormlike larvae that hatch from these eggs reduce the prey to a pulpy mass, presumably allowing nutrients to be more readily absorbed by the plant. The other welcomed guest is a minute mite that also helps break down the insect carcasses. This nutritional fluid is then reabsorbed by the plant.

The plants do catch larger animals, such as the Pacific chorus frog; I have witnessed this among my greenhouse plants many times. It is possible that the frogs are tempted to enter the hood by the insects trapped there, but more likely these amphibians are simply seeking shelter.

A dense stand of *Darlingtonia californica* in a wet meadow in Oregon.

Cobra plants often begin their growth for the season by developing flower buds at the growing point, even in winter. As the weather warms, the single nodding flower arises on a 1- to 3-foot-tall (0.3–0.9 m) stem. The blooms hang upside down, in a fashion similar to those of *Sarracenia*. The sepals are long, pointed, and green; the petals are purplish red and are held together except toward their ends, where they allow for the entrance of pollinating insects.

Cobra plants in live sphagnum at California Carnivores.

Pitchers begin to grow in spring, developing in an upright, rosetted fashion with most of the leaves facing outward. The tallest pitchers are usually the first ones of the season; they can be up to 4 feet (1.2 m) tall in the wild, but usually half this in cultivation. Pitchers produced in summer are usually much shorter than the spring leaves. Grown from seed, maximum-size leaves can take seven to ten years to develop, but the plants are mature and can flower sooner than that.

This plant's most unusual trait is its ability to produce long stolons, or runners, underground. From the tips of the stolons, plants emerge that can reach maturity rather quickly. In the wild, cobras produce dense, impenetrable stands of these runners, sometimes at a great distance from the mother plant. At times these stands seem to run down a seeping slope like a river. In cultivation, these stolons may wrap around the interior of the pot several times before emerging as a baby plant.

Varieties and Cultivars

Cobra plants in the wild as well as cultivated plants can show some variability, but few of these have been mass-produced in cultivation. Some individual plants may have rather enormous, deep red, mustachioed tongues; others are smaller or only tinged with red. Occasionally, plants may be deep red not only along the puffed hood but also along the tubular pitcher body. I am familiar with one plant among many introduced and now naturalized in a bog on the California coast in Mendocino County, in which the insect-luring tongue is entirely lacking and only a residual nub exists.

There has been at least one cultivar registered with the Society and named by Barry Rice. *Darlingtonia* 'Othello' is a plant entirely lacking any red coloration in pitchers or flowers, being anthocyanin free and thus completely green.

CULTIVATION

(See Parts One and Two for further information.)

Cobra plants have a bad reputation for being difficult to grow. This is not far from the truth, but the difficulty can be overcome. You get a head start if you live in a climate with cool summer nights, if you can water your plants frequently, and if you can refrigerate the water (see Watering, opposite).

SOIL RECIPE: Use four parts long-fibered sphagnum moss to one part perlite. Another good mix is one part perlite, lava rock, and/or pumice to four parts peat. The mix should be airy—the rock ingredients will help cool the roots.

CONTAINERS: Plastic pots can be used in a cool environment, but avoid dark colors such as black, which will warm considerably in the sunlight. For best results, use pale-colored, glazed ceramic or terra-cotta clay. Baby cobras with pitchers less than 4 inches (10 cm) long do fine in pots 3 to 4 inches (7.5–10 cm) across. Larger plants need larger pots so stolons can grow around the interior; 12-inch (30.5 cm) pots are suitable for mature plants. The roots can be kept cooler in larger pots, and even small plants will need room for stolon growth. Always use drained containers with holes; never grow cobras in small, undrained pots.

WATERING: Use the tray method, but it is crucial that you continue to water the plant from overhead. Always use cool water. Growers in hot climates should consider refrigerating their water. It often helps to flush the pot with chilly water when temperatures rise. Ice cubes of purified water on a very hot day, placed on the soil surface, can save a cobra's life. Plants originating from rare areas of summer-warm soils, which would not require chilled water, are still uncommon in cultivation. Check with your supplier.

LIGHT: In the wild, cobra plants grow in full sun to partly shaded areas, with best coloration in the sun. In warmer climates, consider having the pots somewhat shaded, perhaps by other containers, if the plants are grown in sun.

continued

Cultivation, *continued*

CLIMATE: A warm temperate to temperate climate, preferably Mediterranean, is needed. Most *Darlingtonia* in cultivation originated from coastal Pacific regions, which is the milder portion of their range. Cool summer nights are preferred—you will have difficulty if your summer nights are very warm. Cobras are very tolerant of frosts. Plants originating from high elevations are more tolerant of frigid winters and hotter summer afternoon air, but still require both cooler soil and night temperature drops into the 50s°F to low 60s°F (10–16°C).

DORMANCY: Cobras require a chilly to cold winter dormancy. Although they are dormant in winter, they can hold some leaves even under a blanket of snow.

OUTDOORS: Generally succeeds well only in Mediterranean climates or high latitudes where summer nights are chilly. Excellent near the coast from Seattle to San Diego or similar coastal climates where summer heat is rare and night temperatures are chilly. Protect pots from hot sun. Water frequently; a turkey baster is helpful to recirculate water from the tray more easily. Drained pots submerged in the ground may also keep the roots cool.

BOG GARDENS: A superb bog candidate if conditions are similar to those under the preceding "Outdoors" description. May succeed in northerly latitudes such as the Great Lakes, New England, or northern Europe if higher elevation plants are used.

WINDOWSILLS: Surprisingly good candidate for partly sunny windowsills. Flush pots often with cool or refrigerated water. Try ice cubes on hot days, or grow in an air-conditioned room. Cool nights are very helpful. Hot sun will warm pots to a dangerous level. Respect dormancy.

TERRARIUMS: Impractical, but young plants may be worth a try in cool terrariums or under grow lights in basements. Flush pots often with chilly water, as recommended in "Windowsills." For winter dormancy, remove the plants and place in a colder environment.

GREENHOUSES: Does well in cold houses, cool houses, and even warm houses, if night temperatures drop below 65°F (18°C) in the summer. Succeeds in cold frames in warm temperate to temperate and Mediterranean climates.

FEEDING: Sow bugs and crickets work well. Dried insects are easier, however.

FERTILIZING: A light foliar feeding once or twice a month during the growing season works well. Use an acidic fertilizer, Orchid 30-10-10, epiphytic, or Maxsea.

TRANSPLANTING: Cobras are best transplanted and divided during dormancy in late winter.

GROOMING: Cobra pitchers can remain in good shape through winter dormancy. Cut away any old brown hoods and pitchers.

PESTS AND DISEASES: Cobras can be attacked by scale, mealybug, and thrips. Aphids can attack newly emerging leaves, and caterpillars love new, soft pitchers before they harden off. Use the pest controls recommended in the Pests and Diseases section, such as a pyrethrin/canola oil treatment. Deer have a peculiar fondness for the puffed heads of *Darlingtonia*. In their native habitat, cobras are sometimes called "deer licks" by the locals. Powdery mildew and similar fungal diseases can cause blotchy purple-red spots on cobra plants, even in wild populations. Use a fungicide such as Serenade or Physan as recommended in the Pests and Diseases section (see page 44).

PROPAGATION

SEED: Seed is a slow but possible way to produce cobras. Pollinate the flowers by hand to ensure good seed set. Separate the petals, and with a tiny paintbrush collect some pollen from the anthers, which you will find on the ceiling of the flower. Deposit this onto the star-shaped stigma, which lies at the bottom of the ovary. Seed is produced two to three months later, as the

continued

The nodding flower of *Darlingtonia.*

Propagation, *continued*

capsule becomes erect and brown, splitting open.

The small seeds are bristly, no doubt to encourage dispersal by furry animals such as deer, bear, and Bigfoot. Store the seed dry in the refrigerator until the following winter. Sow them around February on long-fibered sphagnum. Then stratify the seed for a couple of months, keeping it damp and chilly. As spring approaches, the seed will begin to germinate. The plants will be slow growing. About three years later, the pitchers will be 2 to 4 inches (5–10 cm) long.

DIVISION: You can divide large, mature clumps of *Darlingtonia* periodically. This is best done in winter when the plants are dormant. Make sure each growing point has its own root system. Pamper them with lots of overhead watering as the plants establish themselves.

STOLON CUTTINGS: This is by far the most reliable way to propagate cobras. Stolons can be removed from the mother plant in winter or early spring. The long runners should not be removed until a baby plant is visible at its end. You will notice small roots every 2 to 3 inches (5–7.5 cm) along the length of the runner. Cut the runners into sections, each containing a few roots. I like to lay these horizontally on a bed of sphagnum in a drained seed tray. Keep covered with a propagation dome, and water with cool water frequently. Soon pitchers will develop out of the stolon sections.

TISSUE CULTURE: Cobras can be propagated in vitro through seed.

The Sun Pitchers

(Heliamphora)

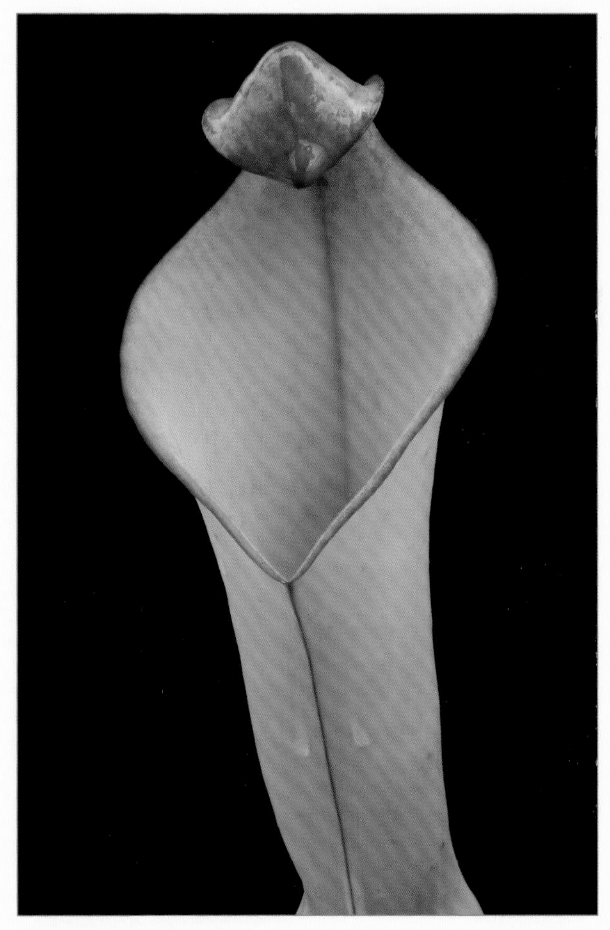

Heliamphora heterodoxa.

Reminiscent of delicately sculpted vases, with tall stems of lovely, lilylike flowers, the sun pitchers of South America have a beautiful simplicity. Related to their cousins in the north, *Sarracenia* and *Darlingtonia*, these are the only pitcher plants of the southern continent, and their native habitat is as exotic and mysterious as the plants themselves.

When Sir Arthur Conan Doyle wrote his famous novel *The Lost World*, he chose as its location the remote and isolated tabletop mountains known as "tepuis." Doyle's prehistoric monsters had survived the millennia on these flat-topped mountains in the Guayana Highlands of Southern Venezuela, Guayana, and northern Brazil. The surrounding savannas are already 4,000 feet (1,219 m) above sea level, a cool tropical climate. The tepuis rise an additional 2,000 to 4,000 feet (609–1,219 m) above this—sheer cliffs and terraces that vanish into the clouds. Although no dinosaurs have yet been discovered, these islands in the sky have many strange species of life on them, and among the most curious are the sun pitchers, *Heliamphora*.

The first species discovered was *H. nutans*, found by the German naturalist R. H. Schomburgk in 1838, on Mt. Roraima. In 1840 the plant was named by G. Bentham, who chose the Greek words *helos* (marsh) and *amphora* (vessel or pitcher). But the similarity to helios, the sun, has given the genus its common name of sun pitcher. It wasn't until almost a century later that a second species was discovered, and in the 1950s and 1970s several more were found as explorations of the many tepuis became more frequent. The knowledge and taxonomy of *Heliamphora* has changed dramatically since the 1990s. Interest in the plants and the use of helicopters to reach the towering tabletop mountains allowed many tepuis to be explored for the first time or reexplored. Among these intrepid scientists and naturalists, several Europeans led the way: Andreas Wistuba, Joachim Nerz, Andreas Fleischman, Stewart McPherson, Thomas Carrow, and Peter Harbarth. They not only discovered new species but also realized that many old collections lying for decades in herbariums had been misidentified. As a result, the species count has risen from a half dozen or so to at least twenty-three, with still more revelations expected to come. This means that there are more species of *Heliamphora* in South America than there are *Sarracenia* in the northern continent.

The tepui mountains of the Guayana Highlands peak among the cool clouds that arise from the humid, tropical landscape below. These clouds condense as they rise, and the result is torrential and frequent rainfall, thunder, and lightning, with frequent high winds. The temperatures at the top can dip near the freezing mark but average 45°F to 52°F (7–11°C) at night to 60°F to 72°F (16–22°C) during the day when the sun peeks through the clouds. The mountains are made of sandstone, and the rains collect in marshes and shallow pools along the surface and end up cascading over the edges into often spectacular waterfalls. Angel Falls, for example, the tallest waterfall in the world, drops almost 3,500 feet (1,067 m) from the top of Mt. Auyan-Tepui in Venezuela.

Kukenan Tepui.

Vegetation is sparse on the mountaintops, a situation ideally suited for carnivores. Since the rains carry away the few minerals that exist, bladderworts and some sundews can often be found growing with the sun pitchers, as can the insect-eating bromeliad *Brocchinia reducta.* There is no real soil for plants to grow in on the tepuis, so these species often grow in the decomposing debris of their own leaves, as well as the short clumping grasses and mosses that often accompany them. Two species have migrated from the mountains to the warmer savannas below but grow in cool water marshes.

Of the species known to exist, the basic structure is very similar. The pitcher leaves arise from a rhizome that is anchored by wiry roots. The pitchers themselves appear to be funnel- or bell-shaped structures that seem nothing more than rolled-up leaves joined by a seam at their front. Although lacking in the fancy ornaments of tongues, hoods, and light windows of their North American cousins, the design of the sun pitchers is nonetheless a rather sophisticated trap.

Adapted to high winds, the pitchers are rather tough yet brittle. Scattered nectar glands appear externally, as a lure to insects. The most prominent structure is a cap on the upper rear of the leaf, known as the nectar spoon. In strong light, these spoons are colorful and well

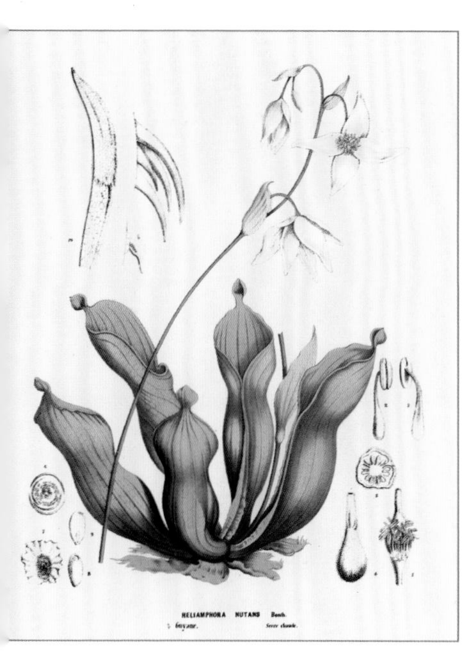
Heliamphora nutans.

developed, hanging over the wide, funnel-shaped mouth like a rudimentary hood. Although the nectar spoon does nothing to prevent the access of copious amounts of rainfall from entering the leaf, it is so positioned as to protect the large glands found in the spoon, which secrete ample amounts of nectar. Nectar glands also exist along the upper interior of the pitcher.

Adrian Slack commented that the small nectar spoon may be a well-adapted design to accommodate no more than one medium-size insect, as he observed two houseflies jostling for a drink of nectar that almost cost the life of one fly.

Once an insect has either slipped off the nectar spoon, or has ventured further inward to reach the interior nectar glands, it finds that the upper, interior walls of most *Heliamphora* pitchers are covered in bristly, downward-pointing hairs, very similar to some species of *Sarracenia*, which make it difficult to climb back out. In some species, the middle section of the interior is completely lacking in hairs, the smooth walls offering little foothold for most insects. These hairs reappear near the bottom of the leaf but are nearly always submerged in the pool of collected rainwater where the insects ultimately fall and drown.

Another curious yet invisible design is a rather narrow slit or pore found about midway along the front seam of the leaf. This allows excess rainwater to trickle free of the pitcher, keeping the water held in the leaf at a constant level.

No drugs have as yet been isolated from the nectar of sun pitchers, nor any digestive enzymes. Apparently, bacteria help dissolve the prey that drown in the rainwater held by each leaf. Also, mosquitoes are known to breed in the pitchers and spiders and even lizards have been seen among them, snatching insects lured by the nectar spoons. Carcasses and feces dropped into the pitchers would no doubt still benefit the plant. Some botanists believe enzymes may yet be found as better methods of research are discovered.

I've seen numerous photos of colorful, bug-eyed frogs sitting in the *Heliamphora* leaves. At California Carnivores, I commonly find Pacific chorus frogs nestled rather comfortably in the well of the leaves, staring in a trance at the nectar spoon above. These frogs are never caught and ingested by the plant. However, it wouldn't surprise me if doing so turns out to be a further adaptation of the plant. The nectar spoon may allow the occasional frog to be literally spoon-fed insects—with a flick of its tongue, the frog can snatch a meal from the plant. The frog's presence may be a further example of "digestion by proxy," as sun pitchers benefit from the excrement the frogs drop into the leaf.

The flowers of *Heliamphora* are showy and beautiful. Tall stems, often blood red, bear a row of large, bell-shaped flowers. The flowers have no petals but have evolved tepals, a structure midway between petals and sepals. The plants usually bloom in winter and spring, and the flowers open progressively along the stem. The anthers and stigmas mature at differing times, thus encouraging cross-pollination by bees. The tepals are usually snow white when they open and gradually turn to pinks and greens over the many weeks that they remain on the stems.

Species

The identification of species can be difficult for several reasons. Most are very similar to each other; their environment both in the wild and in cultivation can greatly affect the appearance of the leaves; they are quite variable; and natural hybrids can also occur. True identification must rely on flower anatomy, which is beyond the scope of this work. Location data using words such as "Massif," "tepui," and "Cerro" basically mean "mountains." When I indicate "lowlands," remember that the Gran Sabana from which the tabletop mountains arise average around 3,000 feet (914 m) in elevation and are therefore somewhat cooler than true tropical climates at sea level.

Here I will discuss first the most common three species in cultivation, followed by some of the more interesting ones, most still rather rare in cultivation.

Heliamphora nutans

The first species, discovered in 1838 at Mt. Roraima and surrounding tepuis, is probably the most common in cultivation. Many of the cultivated plants are descended from lowland populations that are now extinct in the wild due to an accidental 1926 firestorm that was human caused. The curvaceous pitchers are robust, 6 to 12 inches (15–30.5 cm) high, with a bulbous bottom, a thinner waist, a flared mouth, and prominent red helmet-shaped nectar spoons. An easy and rewarding plant for the cool and warm greenhouse or larger terrarium.

Heliamphora heterodoxa

From the foothills of Ptari Tepui and the surrounding Gran Sabana lowlands, it has a helmet-shaped nectar spoon that hangs over the mouth, which in some strange varieties can be elongated and strap shaped. The pitchers are green, outlined in red, to nearly 12 inches (30.5 cm) high. Another robust species from the warmer lowlands that easily takes cooler temperatures as well.

Heliamphora minor

Discovered on Auyan Tepui in the 1930s, the pitchers are most commonly small, averaging 6 inches (15 cm), but occasionally forms can be twice as high. Most plants have tiny nectar spoons, but these can also be larger and bulbous. The pitchers can be green to dark red. Two varieties are recognized: the typical var. *minor* and the bizarre var. *pilosa*, in which the pitcher exterior is covered in fine white hairs.

Heliamphora tatei

Other species were long confused with this one, found on Cerro Duida and surrounding tepuis in 1928. It is famous for slowly forming erect snakelike stems up to 6 feet (1.8 m) long that usually lean on surrounding vegetation. At the top of the stem are two or three active pitchers around 12 inches (30.5 cm) high, green to bronzy, with a large red nectar spoon. Older pitchers turn brown and fall off the stem.

Heliamphora neblinae

Once confused with *H. tatei*, this species was collected in the 1950s on Neblina Massif and other nearby tepuis, but not named until 1978. The stem is usually 1 foot (0.3 m) tall but can occasionally reach 6 feet (1.8 m) if supported by surrounding plants.

Heliamphora ionasi

Discovered in 1952 on Ilu Tepui but not described until 1978, the pitchers can be over 12 inches (30.5 cm) tall. They are widely flared and bell shaped, with a small spoon. The interior bristly hairs emerge from tiny pimplelike bumps.

Heliamphora macdonaldae

Discovered in 1928 on Cerro Duida, the stunning pitchers can top 12 inches (30.5 cm), with a remarkable bright red nectar spoon that is cone shaped, reminding one of an Indian teepee with its flaps open. Also unusual are the bristly hairs that line the pitcher rim; the interior is smooth and hairless, but can sometimes be red veined. Occasionally this species can also form short stems.

Heliamphora folliculata

Discovered in the 1980s and named in 2001, these pitchers come from the remote chain of Los Testigos tepuis. The tubular pitchers reach 12 inches (30.5 cm) tall. Remarkable for its bulbous, hollow spoon that drips sweet nectar into the pitcher.

Heliamphora chimantensis

First collected in 1955 on Chimanta Massif but described by Wistuba and colleagues in 2002. The pitchers can reach 16 inches (40.5 cm) with a long V-shaped frontal slit and an oval, pointed nectar spoon that is held upright. Massive clumps can be 18 feet (5.4 m) across in the wild—no doubt decades, if not centuries old—and the sweet aroma of nectar is noticeable from some distance.

Heliamphora ciliata

First found in 1984, named in 2009 from lowland swamps around Aprada Tepui, the 8-inch (20 cm) pitchers are purple with an overhanging spoon that has a tuft of wiskery hair on its back. Bristly white hairs line the interior.

Heliamphora parva.

Heliamphora parva

Discovered in 1954, named in 2010, from the Neblina Massif. Stem forming, and the tubular pitchers have very large, wavy, flat spoons.

Heliamphora elongata

Discovered in 1984, named in 2004, from Ilu Tepui and surrounding areas. Its habitat is the cold, barren, windswept summits, where temperatures can dip to near freezing. The red-purple pitchers reach 12 inches (30.5 cm) high, and the interior hairs can have a golden-silver shine. In cultivation the temperatures need not be so cold as its native habitat, but should probably still be chilly.

Heliamphora pulchella

The name *pulchella* means "beautiful," which aptly describes this species, known since the 1940s but not named until 2005. Found on the Chimanta Massif and surrounding tepuis, the variable pitchers are short,

Heliamphora pulchella growing in a waterlogged marsh on the Chimanta Massif.

usually around 6 inches (15 cm) tall, and most often the deepest, darkest purple with small spoons. In stunning contrast, the interior of the pitchers is lined with long, silvery, downward-pointing hairs, a devilish trap for insects. Often forming dense clumps, the plants can sometimes be flooded up to the pitcher mouths, but in cultivation keeping this species that wet with stagnant water can rot them. In some varieties the hairs can be sparse or confined to the often waterlogged lower part of the trap.

Heliamphora sarracenioides
This most incredible discovery by Carrow, Wistuba, and Harbarth in 2004 on Ptari Tepui leaves many carnivorous plant growers in speechless awe. It is structurally different from other sun pitchers and reminiscent of its cousins in North America—hence its name and some speculation that it may be a type of "missing link" between the two genera. It is very rare; only a few populations exist. The pitchers can reach 12 inches (30.5 cm) but are usually smaller. They are tubular, somewhat bulbous near the bottom but narrowing toward the top. The sturdy pitchers completely lack a nectar spoon—the back of the pitcher is elongated and upright or curves as a wavy hood over the mouth, tapering to a pointed flap. The pitchers are purple, but the interior, especially along the unusual hood, can be nearly black and is covered in nectar glands.

The bizarre *Heliamphora sarracenioides*—a missing link from the Lost World?

Hybrids

Eleven natural hybrids have been identified in the wild; all of these are of recent discovery or realization. Space does not allow me to cover the details of the plants, but all appear to show the physical combinations of the parent species, and all are simple hybrids between two species growing near each other with the possibility of backcrossing.

Probably all *Heliamphora* will readily cross with each other, and already some artificial simple hybrids have been made, such as *H. heterodoxa* × *minor*, which also shows hybrid vigor. We are only at the very beginning of the history of sun pitcher hybrids in cultivation, but considering the difficulties in pollinating the plants (unlike the ease of crossing *Sarracenia*), this future may unfold slowly.

CULTIVATION

(See Parts One and Two for further details.)

SOIL RECIPE: *Heliamphora* are very tolerant of a variety of wet, open, well-drained soils that are low in nutrients. An excellent mix is three parts long-fibered New Zealand sphagnum moss, one part lava rock, and one part perlite. Alternatively, you can use about half sphagnum peat moss with an equal combination of lava rock, pumice, perlite, and fine orchard bark. Live sphagnum can make an attractive topdressing for larger plants but will certainly overwhelm young ones, which is probably why *Heliamphora* avoid growing in sphagnum, which is found on some tepuis.

CONTAINERS: Plastic or glazed ceramics that have drainage holes. Four-inch (10 cm) pots for young plants, 6- to 8-inch (15–20 cm) pots for mature plants.

WATERING: Use a shallow tray method with frequent overhead watering with cool, purified water. If watered daily, the pots need not sit in trays or saucers. Keep the soil permanently wet.

LIGHT: High light levels or sunny conditions are best for healthy, colorful plants. Avoid overheating pots.

CLIMATE: Most are considered highland tropicals, and do best with night temperatures between 45°F and 60°F (7–16°C) and day temperatures between 60°F and 80°F (16–27°C). Extremes briefly tolerated are 35°F to 90°F (2–32°C), but cooler day temperatures are preferred. High humidity is a requirement, as well as good air circulation.

DORMANCY: Sun pitchers require no dormant period.

OUTDOORS: Very doubtful, unless you live in highland tropics. They might succeed on a cool, foggy, frost-free coastline like California, but would need protection from salty spray and hot, dry offshore winds.

BOG GARDENS: Generally not suitable, for reasons cited under "Outdoors."

WINDOWSILLS: Very doubtful, due to the plants' high humidity and sunlight requirements.

TERRARIUMS: Excellent in cooler terrariums under grow lights. If you live in a warm climate, terrariums in basements or air-conditioned rooms work well. Refreezable ice packs can help cool terrariums overnight. Mist your plants often.

GREENHOUSES: Excellent for the cool house and warm house, near swamp coolers or under mist systems. Try not to exceed 85°F (29°C). A drop in night temperature is a requirement.

FEEDING: Young plants will eat fruit flies and baby crickets; larger plants will enjoy sow or pill bugs, crickets, mealworms, and dried insects.

FERTILIZING: *Heliamphora* respond very well to foliar feeding. Apply this once or twice monthly and the plants will grow fairly fast and flower regularly.

TRANSPLANTING: Easy, and best done in winter and spring. A soaking in vitamin B1 solution can be helpful. Be gentle when transplanting, as pitchers can be brittle and easily damaged.

GROOMING: Trim away the old pitchers as they turn brown.

continued

PESTS AND DISEASES: Rarely attacked by scale. Fungus might be a problem in stagnant air and low light levels. Slack mentions that copper-based fungicides are lethal.

PROPAGATION

SEED: Observation on the tepuis has indicated that it is bees that pollinate *Heliamphora* flowers, and that similarly to other species of plants, vibration is needed to release pollen.

The several flowers on a scape will open one at a time over the course of a few weeks. When a flower opens, the female stigma is receptive for a few days, but the male pollen is not mature until after the stigma loses its receptivity. Therefore, flowers must be cross-pollinated. When examining the flower of a sun pitcher plant, you will see a cluster of tubelike pollen sacs—the anthers. Rising from the center of this is a short green stalk, the tip of which is the female stigma. The stigma turns brown when it is no longer receptive.

The flowers of *Heliamphora nutans*—a pleasure any time of the year.

The tuning fork method works well to release pollen. Hold a piece of paper below the tepals and touch the pollen sacs with the vibrating fork. A small, dustlike amount of pollen will be deposited if the pollen is mature. Collect this with a small paintbrush and dab it onto the green stigma of another, newly opened flower. If pollination is successful, the ovary will swell, turn brown, and release its seed within two to four months.

The seed need no stratification and can be lightly scattered on milled sphagnum or their preferred soil mix. Keep humid and in bright light, and germination should occur in several weeks. A seedling will take several years to reach maturity.

Heliamphora may also be cross-pollinated among their species with ease, thus producing hybrids. Some hybrids already produced include *H. nutans* x *minor*, *H. heterodoxa* x *minor*, and *H. nutans* x *heterodoxa*, among others. These crosses have the combined characteristics of both parents and usually have hybrid vigor.

DIVISION: Sun pitchers have the pleasant habit of vigorously producing offshoots when grown well, thus producing handsome clumps over time. These can be easily divided when necessary.

I prefer to do this during winter and spring. Remove the plant from its pot and wash away the soil. You should be able to clearly distinguish the various growing points along the rhizome. Gently snap or cut these apart. Try to ensure that each growing point has its own root system. It can be helpful to soak in a vitamin B1 solution, such as SUPERthrive, and repot. About two weeks later, apply a light mist of fertilizer. Remove any flower stalks from a divided plant, because these will exhaust its recovery.

TISSUE CULTURE: *Heliamphora* have become much more common and affordable through the success of tissue culture. Plants may be introduced into flask through the use of sterilized seed.

The Sundews

(Drosera)

"Evil little things they are, with their carnivorous habit. One wonders what crime the past lives of *Drosera* can have held, that now their race should be compelled to draw so ominous and unpleasant a world of murder and fraud. When will Sundews be free of the burden, through some self-sacrificing individual plant who shall starve to death rather than take life, and so redeem his race into happier paths of peace and virtue?"

—Reginald Farrar, *Alpine and Bog Plants*, 1908

Sundews at sunset.

If an insect ever evolved the brains to write a horror novel, the monster in that novel would probably be a sundew.

Sundews are innocent looking and pretty, their delicate leaves sparkling with the promise of sweet nectar, but the foolish insect curious enough to give a sundew the slightest touch will suddenly find itself caught in a living nightmare. Doomed to a horrible death, the insect may struggle for a blessed few minutes or suffer for untold hours as it tries to break free of ensnaring, suffocating glue, grasping tentacles, and burning acids and enzymes; meanwhile, its precious bodily fluids are slowly sucked dry. Mother Nature hopefully had psychiatric care after she designed the sundews.

The *Drosera* are probably the most diverse genus of carnivorous plants in the world. There are somewhere around two hundred species found on almost every continent of earth. Sundews can be found in Canada, Alaska, and Siberia. They are denizens of bogs and swamps in much of Europe and North America. They lurk in tropical places such as Brazil and Queensland. And they haunt the southernmost regions of New Zealand and South America.

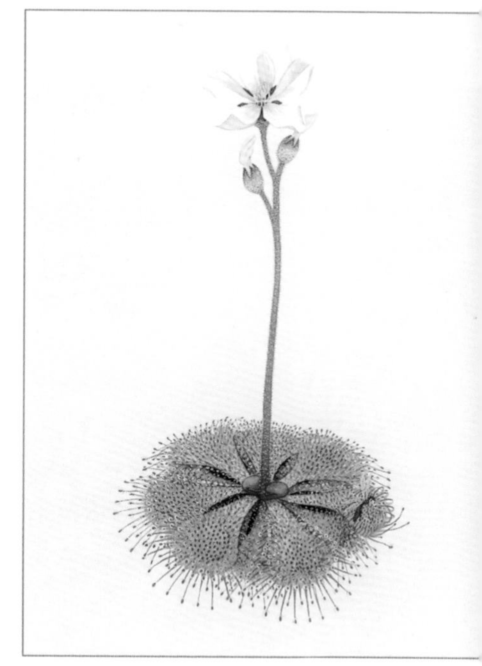

Drosera brevifolia.

Sundews can be as tiny as a penny or as large as a small bush. Their tentacle-covered leaves come in a wide and imaginative variety of design: circular leaves, wedge-shaped leaves, leaves that are peltate or linear or as filiform as a thin blade of grass. Their leaves may be strap shaped, oval, or forked and branching like a dewy fern or lethal spiderweb.

Their adaptation to diverse climates is also rather inventive. While all sundews need wet, low-nutrient soils to grow, some have adapted to survive hot, dry summers and grow only during the cool winter rains. Others are found in bogs that are frozen much of the year. Still others grow in Mediterranean climates where temperatures may drop to near freezing at night and rise dramatically during the day.

A pleasant fact about sundews is that they often enjoy the same hunting grounds as many other carnivorous plants. Visit flytraps and American pitcher plants in North Carolina and you'll no doubt step on sundews in the process. Go to the tepui mountains in Venezuela to photograph the sun pitchers and *Brocchinea*, and sundews will probably be in the picture. Sundews grow with *Cephalotus* in Australia, *Darlingtonia* in Oregon, and *Nepenthes* in Borneo. In fact, whenever you find wet, low-nutrient, acidic soils, chances are sundews, the ubiquitous carnivorous plant, will be there.

Sundews caught the attention of European naturalists early on. Henry Lyte, an early British botanist, wrote of the peculiar nature of the plants in his "New Herbal" in 1578. Commenting (in Old English) on how the plants appeared to increase their dew the hotter the sun became, rather than the opposite, he mentions their Latin name to be *ros solis*, "which is to say in English, the Dewe of the Sonne, or Sonnedew." The word *Drosera* in Greek means "dewy."

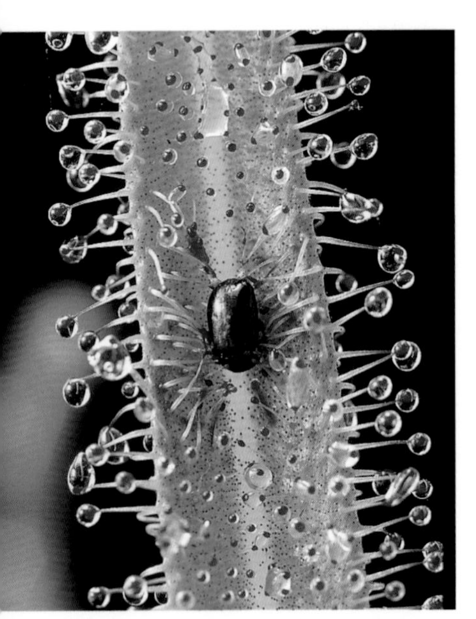

There's no escape for this beetle on a leaf of *Drosera regia*.

In 1791, Charles Darwin's grandfather, Erasmus, thought the dew produced by the plants protected them from predators. That same year, William Bartram concluded insects were purposely caught, but for reasons he could not surmise. A German botanist, Dr. W. A. Roth, was the first to notice that the tentacles of sundews actually closed around prey in 1779.

But it wasn't until Charles Darwin himself took up the study of sundews that they were ultimately proven to be carnivorous. Darwin conducted countless experiments on the plants for many years and published the findings in his 1875 *Insectivorous Plants*. The bulk of the book was on his studies of the common round-leaved sundew, *Drosera rotundifolia*. At one point, Darwin wrote to his botanist friend Charles Lyell that he cared more about sundews than about the origin of all the species of life on earth. He also described them as being like animals in disguise. He found that the plants were more sensitive to taste and touch than any animal species he had studied.

Although the size and shape of *Drosera* leaves greatly vary, their insidious design and function are similar among all the species.

The leaves are generally flat, and their uppermost surface is covered with hundreds of stalked glands or tentacles. The tentacles are hairlike filaments, at the end of which is a small, usually reddish, gland. Most of these glands produce a tiny drop of dew: a clear, gluey mucilage that is extremely sticky. On some species of sundews, the glands on the outer edge of the leaf do not produce glue; they are called retentive glands, for reasons we shall see.

A small flying insect may catch sight of the glistening droplets and mistake them for a flower's sweet nectar. It alights on the leaf and immediately panics. The "nectar" is glue, and the insect struggles to free itself. Instead, its legs and wings come into contact with more of the sticky drops, and the more it thrashes the more mired it becomes. It may lose a leg or a wing in the fight, or it may pull partly free of one leaf of the sundew only to be caught by the tentacles of another.

Then a second lethal action comes into play. Within moments, some of the tentacles begin to move. If the insect struggles toward the edge of the leaf, the retentive glands along the edge begin to curl inward, blocking the panicking insect's escape. If the insect happens to have been caught by some outermost glands, those tentacles quickly begin to move, carrying the victim toward the center of the leaf, where dozens more glands await it.

Usually the prey of sundews suffocate when the breathing holes along their abdomens are covered in glue. But larger insects, such as crane flies, are sometimes caught by their long, thin legs, their bodies dangling helplessly below the leaf. These victims usually die of exhaustion or starvation while the plant feeds on only the body parts it has caught.

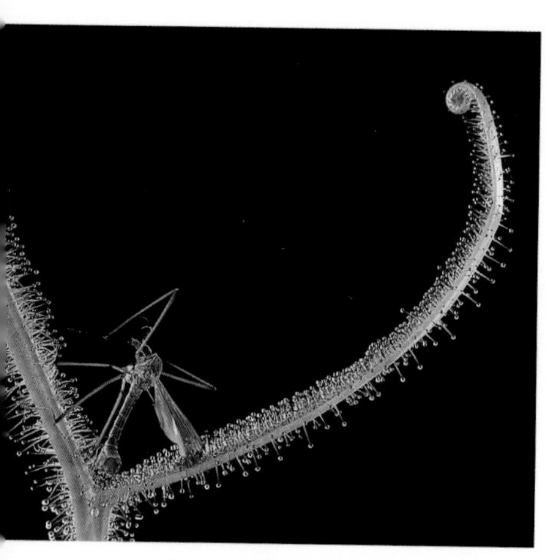

A crane fly realizes its mistake in landing on a staghorn sundew.

Tentacles have seized the crane fly's abdomen, and digestion begins.

Occasionally, larger insects escape—minus a few legs—only to die later. Sometimes they are finished off by spiders, which enjoy spinning webs around sundew plants. Spiders, too, are often caught, usually the consequence of their own greed, as they attempt to snatch a meal rightfully belonging to the sundew that trapped it.

Further movement in some *Drosera* species can be rather dramatic, although this movement typically enhances digestion rather than capture. Many sundews have the capacity to enfold their leaves around their prey. Varieties of sundews with thin, filiform leaves, or leaves that fork or branch, rarely have this ability; they instead rely strictly on the movement of tentacles for feeding. Others, with circular or strap-shaped leaves, may close around their meal like a clenched fist or jelly roll. Still others, like *Drosera regia*, may tie their leaves into knots around a larger meal.

The object is apparently to get as many glands as possible into contact with the prey. For once an insect is caught, the highly developed glands begin to secrete a complex juice of enzymes and acids that rapidly covers the insect's body. This fluid can at times be rather copious, dripping down the leaf. In a matter of hours or days, the digestive juices liquefy the softer parts of the insect. The glands then begin to reabsorb this nutritious, mineral-laden soup.

The movement of sundew tentacles and leaves is caused by a process similar to that of the Venus flytrap, the sundew's relative. It is an amazing growth process. Cells along one side of a tentacle will grow and stretch, and the unequal length of the cells causes the tentacle to bend. Exactly how this occurs and how electrical signals notify nearby glands to start moving, even when an insect is some distance away, is still not clearly understood.

The speed at which sundews move is also variable and differs depending on factors such as temperature, age of the leaf, species of plant, how much the insect struggles, and its nutrient worth. The tentacles won't react to a pine needle falling on them, but will slowly begin to move if they're repeatedly teased with a toothpick. If you place a bit of cheese or chocolate on a leaf, initial movement may be slow, but within a day or two the leaf may be curled dramatically and drooling digestive juices. Typically, in about twenty minutes a number of tentacles may pin an insect to the center of the leaf, while it may take twenty-four hours for a leaf to completely curl around its prey.

Several sundews have retentive glands that move so rapidly the results are startling. Species such as *Drosera pauciflora, D. burmanni*, and *D. scorpiodes* can move 180 degrees within one minute, quickly securing an insect before it knows what has happened. And one tiny species, *D. glanduligera*, has marginal retentive glands called "snap tentacles" that can secure a victim in a fraction of a second.

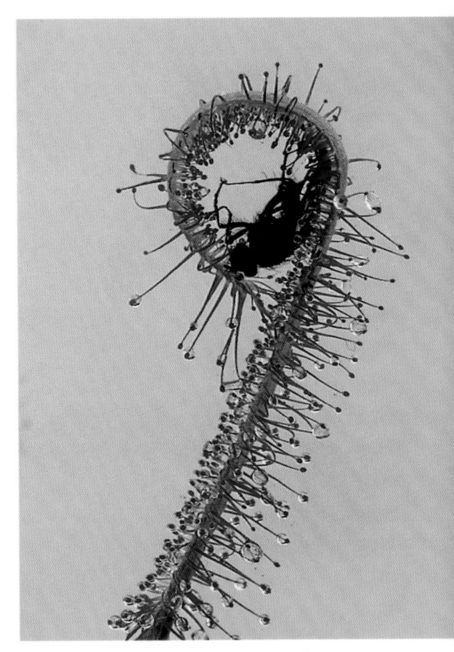

The movement of cape sundews can be rather dramatic, as seen around this helpless fly.

After digestion, the tentacles usually dry up and return to their normal upright position. The shriveled carcass falls from the leaf, is washed off by rain, or simply remains as what should be a dire warning to future prey. Dew drops reappear, and the leaf is ready for another meal. Sometimes older leaves remain curled up even as the leaf turns brown and dies. During their active growth cycle, most sundews continually replace older leaves with new ones.

The flowers of sundews can be as varied as their leaves, and in some cases they are rather spectacular. Almost all *Drosera* flowers are five-petaled, circular, and more or less flat-faced, with the female stigma surrounded by male anthers. As a result, many self-pollinate and produce numerous seed. Others, however, never produce seed because they need cross-pollination, and some varieties in cultivation are almost all of one genetic clone. The predominant petal colors of sundews are whites and pinks, but in some groups, such as pygmy and tuberous sundews, the colors may range from orange to red to yellow to violet. Also, the inflorescence of sundews is quite variable, and while most produce tall flower stalks with rows of blooms, others may be clustered or branching. Further, the size of the flowers can be rather dramatic, and in some unusual species the flowers may be as large as or larger than the plant itself.

Due to the wide-ranging growth habits and habitats of sundews, this chapter is arranged somewhat differently from the others. First, I will discuss cultivation techniques in the most general terms. I will then review the grouping of plants by their characteristics and habitats, including descriptions of representative species and particulars about their cultivation. This will be followed by information on their propagation.

CULTIVATION

(See Parts One and Two for further details.)

SOIL RECIPE: Most species do well in a mix of three parts sphagnum peat to one part sand or perlite. Some species, such as tuberous sundews, prefer a recipe of at least half sand to half peat. A few do best in long-fibered sphagnum.

CONTAINERS: Plastic containers with drainage holes are most suitable. Many species do well in undrained containers of plastic, glass, or glazed ceramic. Small rosetted sundews do well in 3- to 4-inch (7.5–10 cm) pots; larger ones like forked sundews can be grown in 6- to 10-inch (15–25 cm) pots.

WATERING: Most sundews thrive in the tray method, which keeps the soil permanently wet. A few prefer to be waterlogged. Winter-growing species require periods of complete summer dormancy, at which time the soil has to be dried out.

LIGHT: Almost all sundews require sunny to partly sunny conditions. Very few prefer full shade.

CLIMATE: Because sundews grow worldwide, they come from varied climates: cold temperate, temperate, warm temperate, subtropical, tropical, and Mediterranean. Their preferred climates will be indicated under the species descriptions beginning on page 161.

DORMANCY: Many sundews grow year-round with no dormancy requirements. Others require cold winter dormancy or dry summer dormancy.

OUTDOORS: Sundews thrive outdoors in climates similar to their native habitats. The widest variety can be grown outdoors in warm temperate, subtropical, and Mediterranean climates.

BOG GARDENS: Many sundews are ideal for bog gardens. Choose those suitable for your climate. Avoid most species requiring a dry summer dormancy.

WINDOWSILLS: Many sundews, such as cape sundews and rosetted subtropical species, thrive on sunny and humid windowsills. Your house's conditions will best determine suitable varieties. Surprisingly, even spectacular giants such as *Drosera regia* and *D. multifida* can be happy on windowsills if provided with much direct sunshine and humid air.

TERRARIUMS: Due to their preference for sun, most sundews thrive in well-lighted tanks. Fluorescent bulbs are best. The closer to the lights, the more colorful and robust the plants will be. Only tropical and subtropical sundews should be grown year-round in tanks. Plants with seasonal cold winter or dry summer dormancy will need to be removed for their rest period.

GREENHOUSES: Different greenhouse types will suit different types of sundews. Almost all thrive in warm houses and cool houses, where perhaps 80 percent of all *Drosera* species will be happy.

FEEDING: Sundews readily accept gnats, fruit flies, or small ants. Larger sundews will feed on houseflies, spiders, moths, and the like. Strong insects

continued

such as crickets are generally unsuitable. Dried insects and pinhead-size bits of hard-boiled egg, cheese, chocolate, and powdered milk are also accepted.

FERTILIZING

Not necessary if the plants catch many insects, but most species benefit from monthly foliar feeding with a fertilizer diluted to about a quarter of its normal strength or much less. Never apply fertilizer to *Drosera schizandra, D. prolifera,* or *D. adelae,* unless it is Maxsea.

TRANSPLANTING: Sundews can be happy in a container for many years. They are best transplanted when in a dormant state or in early seasonal growth.

GROOMING: Old dead leaves can be trimmed away, but some growers like the old skirt of leaves on some stem-forming plants.

PESTS AND DISEASES: Aphids are the primary pest. Use a suitable insecticide such as pyrethrin/canola oil sprayed mostly on the newly emerging leaves or flower stalks where aphids are likely to attack. Some insecticides, such as Sevin, may damage existing leaves of some sundews, but new growth will be normal. Flea collars in small enclosures work well. Some sundews can be attacked by fungus during winterlike conditions of low light and cool, dank temperatures. Apply fungicide or heavily sprinkle pure water over the plant frequently or allow rain to wash away fungus and spores. Never use soap-based insecticides on sundews.

Cape Sundews

From the cape of South Africa comes this marvelous sundew, *Drosera capensis*, a variable species that offers everything the plant lover could wish for. Cape sundews are large, handsome plants that are very easy to grow. They produce scores of showy flowers on tall stems and are so simple to propagate that they often become a weed in collections. Their leaves move rather dramatically, and they are tolerant of a wide range of growing conditions. These are by far the most entertaining and popular of the *Drosera*, and a perfect beginner's plant.

Cape sundews grow best in a temperature range of 40°F to 80°F (4–27°C). They are very tolerant of extremes from freezing to 100°F (0–38°C), but only for brief periods. The plants grow year-round and have no dormancy requirements. If temperatures drop below freezing, the crowns will die away, but the plants return from their roots and stems when conditions improve. Roots have been known to survive brief temperature drops to 15°F (-9°C).

Cape sundews will repeatedly send up tall flower stalks from spring through autumn. Each stalk can hold up to several dozen flowers, one bloom opening every day or so. The flowers, up to 3/4 inch (2 cm) in diameter, usually self-pollinate and can produce copious quantities of seed that will readily germinate on any damp, peaty soil. The plants are also easy to reproduce via leaf and root cuttings. Continuous flowering can have an exhausting effect on the plants, resulting in smaller crowns by late summer, but the plants recuperate fully when blooming stops. For larger crowns, remove the scape as it develops; this will also prevent rampant seedling growth.

Cape sundews thrive in cold houses, cool houses, and warm houses; terrariums under grow lights; sunny and humid windowsills; and outdoors as potted plants or in bog gardens in warm temperate, subtropical, and Mediterranean climates.

Drosera capensis 'Wide Leaf'

This plant gradually forms scrambling stems several inches (15 cm) long with a cluster of leaves at the apex. The green leaves are strap shaped at the end of wide, lengthy petioles and are covered with bright red tentacles. Individual leaves are 3 to 6 inches (7.5–15 cm) long or larger, resulting in crowns that can reach over a foot (0.3 m) across. This form can be pruned back by cutting off the crown of leaves in late winter or spring. Shoots will develop from the remaining stem, eventually producing

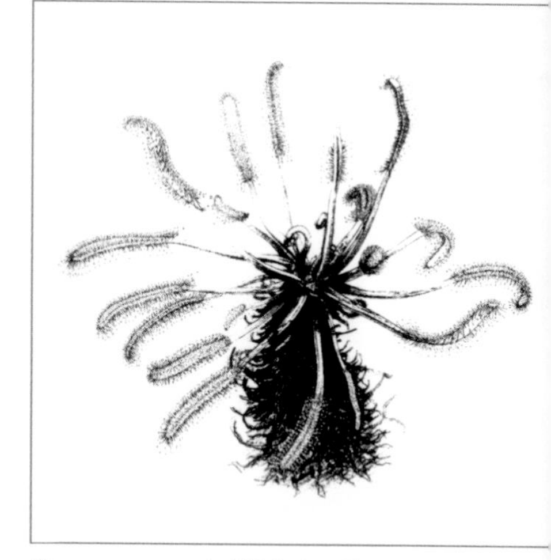

Drosera capensis 'Wide Leaf'.

Cape sundews, from left to right: 'Wide', 'Narrow', 'Red', and 'Albino'.

large, multiheaded, bushy plants. The bright pink flowers, on stalks 1 to 2 feet (0.3–0.6 m) tall, produce enormous quantities of seed.

Drosera capensis 'Narrow Leaf'

Similar to 'Wide Leaf' in almost all respects, except that the leaves and petioles are rather narrow, about 1/4 inch (6 mm) in diameter.

Drosera capensis 'Albino'

This very pretty form is similar to the narrow-leaved variety, except that the flowers are white and the tentacles transparent with pale pink glands, giving the plants a ghostly appearance. Stems may gradually form over time. Also commonly known as the 'Alba' form.

The flower of *Drosera capensis*.

Drosera capensis 'Red'

This stunning variety is entirely reddish maroon, with deep pink flowers. In all other respects, it is similar to the narrow-leaved form. This variety requires lots of sun or very high light levels to maintain its rich coloration, which may fade in winter or when the plants are grown too far from grow lights.

Rosetted Subtropical Sundews

Many sundews from various parts of the world fall into this category, which for the sake of convenience I will call "subtropical," although they can also be found in warm temperate, Mediterranean, and tropical climates.

The basic form of most of these *Drosera* is a low-growing plant with leaves that fan out into a rosettelike pattern. Typically their leaves are pressed flat to the ground, although some may hold their foliage in a more upright fashion, or even form a short stem, with old, dead leaves forming a skirt around their base. Usually these rosetted species average 1 to 3 inches (2.5–7.5 cm) in diameter, although a few may be smaller or larger.

Rosetted subtropical sundews are all pretty plants, and a few are spectacular. They are usually as easy to grow as cape sundews, and many will also spread like weeds from ample seed production. A few are short-lived perennials, meaning that older rosettes die off after a couple of years and the plant reproduces by seed. Almost all can easily be propagated by leaf cuttings. Some have thick roots also useful for propagation.

With a very few exceptions, this class of *Drosera* grows throughout the year without any dormancy requirements. While most grow in frost-free climates, many are tolerant of brief freezes into the 20s°F (-7°C). The rosettes will die during these light frosts, but usually new plants come up from the thick, protected roots. For best results, grow them in frost-free environments.

Rosetted sundews produce flower stalks throughout much of the warmer months of the year. The stalks may be 3 to 5 inches (7.5–12.5 cm) tall, but in some plants may reach 1 or 2 feet (0.3–0.6 m), quite tall for such small plants. In some, a scape of flowers opens, one at a time, over several weeks. These flowers are typically small, barely 1/4 inch (6 mm) across, but in some can be quite showy and large. A few exceptional species have very large blooms, almost as big as the plant itself. While many self-pollinate and produce seed, others never do and must be propagated from leaves or roots. The flowers of rosetted subtropicals are invariably shades of pink or white.

As a group, these sundews perform extremely well in cool and warm greenhouses. Many are exceptional plants for a humid and sunny windowsill. They are also great for the terrarium under grow lights, but for best results should be no more than 6 to 10 inches (15–25 cm) from fluorescent bulbs. They are great outdoor plants if you live in an almost frost-free climate.

Some of these species look so similar to each other that even longtime growers have difficulty telling them apart. Also, new varieties are still being found in countries such as South Africa, and these plants may be circulated under nicknames for many years. For many growers, pedigree is not quite as important as simply having a handsome rosetted sundew to look at!

One of the most popular rosetted sundews, *Drosera aliciae*, from South Africa.

Drosera aliciae

Possibly the most popular rosetted species, the Alice sundew comes from the Cape Province of South Africa. The handsome rosettes are up to 3 inches (7.5 cm) across, producing a mound of pale green, wedge-shaped leaves with crimson tentacles. The leaves curl dramatically around larger prey. The substantial flowers are deep pink, around 1/2 inch (1.3 cm) in diameter, and on stalks over 1 foot (0.3 m) high. This species often produces offshoots from its roots, eventually producing attractive clumps of plants. It also returns well from light frosts. Easily propagated from abundant seed or leaf and root cuttings.

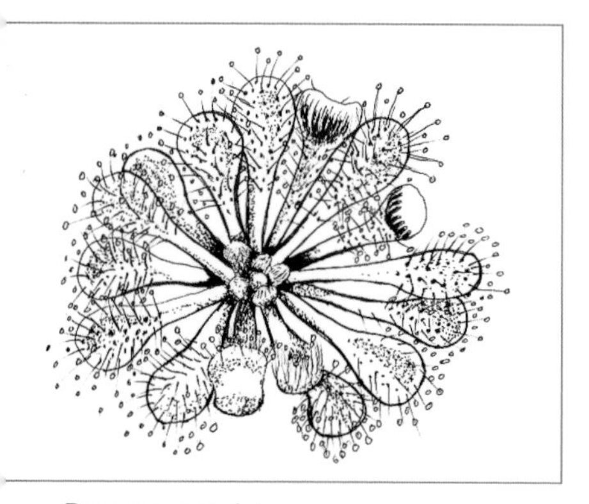

Drosera spatulata.

Drosera spatulata

This widespread and variable species is found from southern Japan and Southeast Asia to Australia and New Zealand. Many forms are grown, usually known by their location. The rosettes are flat, with spoon- to wedge-shaped leaves typically 1 to 2 inches (2.5–5 cm) across. In good light the entire plant can be flushed red. The small flowers are pink or white and usually self-seed. Among the most popular, the 'Kanto' form from Japan has narrow wedge-shaped leaves, similar to forms from eastern Australia. 'Ruby Slippers' has small, dark red rosettes to 1 inch (2.5 cm) in strong light, while 'Tamlin' can be twice as large and orange-red, similar to 'Fraser Island' forms of eastern Australia. Two varieties have been named: var. *gympiensis* from Queensland, with hairy sepals on the flowers, and var. *bakoensis* from Bako National Park in Sarawak, a tiny plant under 1 inch (2.5 cm) with round leaves and a few pale pink flowers. What was once called the 'Kansai' form from Japan is now known as *D. tokaiensis*, a hybrid between *D. spatulata* × *rotundifolia*. While most forms of *D. spatulata* are considered tropical to subtropical, many survive brief freezes or return from roots or seed.

Drosera spatulata var. *gympiensis*, from Queensland, Australia.

Drosera tokaiensis, from Japan.

Drosera capillaris

The pink sundew is very common in the southeastern United States from Virginia to Florida to Texas. Forms are also known from Central and South America. The rosettes are usually 1 to 2 inches (2.5–5 cm) across with almost circular to sometimes spoon-shaped leaves, often reddish. Flowers are usually pink. An 'Alba' form is all green

Drosera capillaris.

with white flowers. This species usually performs as an annual or short-lived perennial, returning from seed after hard frosts. It is very commonly seen growing with other carnivores in the southeast, such as *Sarracenia, Dionaea*, and *Pinguicula*.

Drosera brevifolia

Uncommon in collections, this tiny sundew has a range similar to *Drosera capillaris*. It looks like a miniature *Drosera spatulata* 'Kanto', with bright red, wedge-shaped leaves. The whole plant is barely 1/2 inch (1.3 cm) across, and very pretty in dense colonies. Short-lived, the plants often die away in winter, returning from seed. A story I've enjoyed is how this diminutive sundew has sometimes been found growing in the damp cracks of sidewalks in Houston, Texas, during the wetter parts of the year!

One of the finest of recent discoveries, *Drosera slackii*. Discovered by Frank Woodvine, it was named by Martin Cheek after Adrian Slack, who introduced it into cultivation.

Drosera slackii

This magnificent *Drosera* was discovered in the early 1980s in the coastal mountains of the cape of South Africa and named for Adrian Slack, who introduced it to cultivation. The beautiful rosettes can be large, from 2 to 3 inches (5–7.5 cm) across, and can form a short stem draped in dead leaves. Deep crimson, the leaf blades are almost circular with broad petioles, and they move dramatically over prey. The deep pinkish violet flowers are more than 1 inch (2.5 cm) in diameter. They rarely produce seed, but the plant is easily propagated from roots and leaves and is slowly clump forming. They can do very well on windowsills.

Drosera cuneifolia

Another handsome South African sundew that, with its wide petioles, reminds me of *Drosera slackii*. But the leaf blade is even wider and with rounded corners. The leaves are bright green with pink tentacles.

Drosera dielsiana

A small sundew from South Africa, this plant was named after botanist Ludwig Diels, who wrote the first monograph on sundews early in the twentieth century. The roundish leaf blades have a fairly wide petiole, and the flowers are pink. *D. natalensis*, also from South Africa, is nearly identical except for floral differences.

Drosera hamiltonii

This peculiar sundew grows with *Cephalotus* in southwestern Australia. The flat rosettes have pale, olive brown leaves that are shaped like rounded paddles and sparsely covered in red tentacles. It flowers infrequently; the blooms are purplish and over $1^{1}/_{2}$ inches (3.8 cm) across, almost as large as the rosette itself. The flowers put on quite a show, but do not seed; the plant propagates from its roots and leaves. A winter chill can induce spring flowering.

Drosera hamiltonii.

Drosera collinsiae

A very pretty sundew, this South African species has upright, spoon-shaped leaves on long, thin petioles. The leaves are bright green with red tentacles. This species has a brief winter dormancy, and the pink flowers usually self-seed.

Drosera burmanni

This curious species is a tropical annual that grows in northern Australia and Southeast Asia. It lives only for a few months during the warm rainy season, seeds prolifically, then dies off when the soil dries out, returning from seed when the rains return. In cultivation,

The elongated retentive glands of *Drosera burmanni* are among the fastest moving in the genus.

it is easy to grow during any stretch of warm weather. The plants are about 1 inch (2.5 cm) across, with deeply dished, wedge-shaped leaves, usually green or pinkish, and small white flowers. After flowering, the plant dies, so be sure to collect the tiny seed. The seed will also tolerate a light frost. Most amazing are the long retentive glands, which can close around prey in less than one minute. The recently rediscovered *D. sessilifolia* from Brazil is no doubt related and nearly identical.

Drosera ascendens.

Drosera villosa, D. ascendens, D. graomogolensis, **and** *D. schwackei*

These four South American sundews are rather similar, with arching, strap-shaped leaves. Depending on the species, they can be 2 to 5 inches (5–12.5 cm) across. Considered highland tropicals, they do best with warm days and somewhat cooler nights around 50°F to 60°F (10–16°C).

Drosera schwackei, growing in quartz sand.

Drosera capensis × spatulata

This handsome, compact hybrid has strap-shaped leaves with rosettes to 3 inches (7.5 cm) across, bright green with red tentacles. They usually go into a semi-dormant state during winter chills and can return from their roots after light freezes. The pink flowers are sterile, but it can be easily reproduced through leaf cuttings.

Drosera glabripes

From South Africa, this sundew can form long, scrambling stems, with the rosette of green, spoon-shaped leaves at the apex. The rosette averages 3 inches (7.5 cm) across. This species requires rather chilly winter temperatures when it slows its growth, but avoid frost.

Drosera venusta

A robust South African species with rosettes 2 to 4 inches (5–10 cm) across, this recent discovery is reminiscent of *Drosera aliciae.* The long, oval-shaped leaves are semierect and golden green, which sets off the bright red tentacles rather well. The flowers are pink on tall wiry stems. *D. coccicaulis*, also South African, is very similar.

Temperate Sundews

Sundews that grow in climates that experience various amounts of cold weather in winter survive frost and snow by dying down to buds called "hibernacula" (from the word "hibernate"). Some of these species have wide natural ranges, and their length of dormancy and cold tolerance is based on where the plant or seed originated. You should therefore pay attention to location data, if available, when considering which forms of these species you wish to grow.

For example, one species, *Drosera intermedia*, is actually pan climatic, growing from Canada south all the way into South America. Plants originating from Wisconsin are cold temperate, may experience a dormancy lasting six or seven months, and can survive temperatures well below 0°F (-18°C). Representatives of *D. intermedia* from Louisiana would be considered warm temperate, being dormant perhaps three to four months and surviving only light frost and brief freezes. *D. intermedia* from Cuba are tropical, having no dormancy and an inability to withstand freezing temperatures.

Most of these species are easy to grow if you allow them the dormancy they require. If you live in a warmer climate, you can always remove the hibernaculum once the plant dies down, and refrigerate it over winter in an airtight plastic bag with a few strands of damp sphagnum moss. When dormant, these plants usually lose their roots as well as leaves, and this is the best time to transplant them. They are easy to propagate from leaf cuttings and seed. Most do best outdoors, or in cold and cool greenhouses.

Drosera rotundifolia

This is the plant made famous by Charles Darwin's tireless and hideous experiments on it. The round-leaved sundew is a common inhabitant of almost every sphagnum bog found in the northern latitudes. Its range is

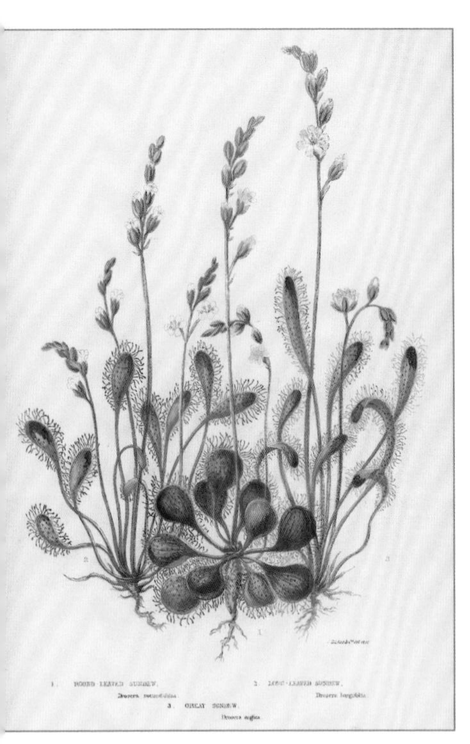

Left to right: *Drosera anglica, D. rotundifolia,* and *D. linearis,* which was once called *D. longifolia.*

from Alaska through much of North America as well as Europe and Asia. The rosettes average 3 inches (7.5 cm) in diameter, with dished oval leaves at the end of long, thin petioles. The flowers are white. Specimens from Northern California, Oregon, and the New Jersey Pine Barrens can have rosettes up to 5 inches (12.5 cm) across, with dime-size leaves. In cultivation, the species grows best in long-fibered sphagnum.

Drosera anglica

The "English sundew" grows in bog habitats in Japan, North America (including Alaska), and Europe. Similar to *Drosera rotundifolia* and often growing side by side with it, *D. anglica* has long, paddle-shaped leaves held more upright. An unusual colony is found in Hawaii on the island of Kauai. These plants are tropical, smaller in stature, and do not go dormant, making them ideal for terrariums.

Strangely enough, evidence suggests *Drosera anglica* is a fertile hybrid between *Drosera rotundifolia* and *Drosera linearis* (described shortly). The mystery is why *Drosera anglica* is so widespread while *Drosera linearis* is presently restricted in its range.

Drosera × obovata

The natural hybrid between *Drosera rotundifolia* and *D. anglica* can be rather robust and handsome, looking intermediary between the two. Sterile, it must be propagated by leaf cuttings.

Drosera intermedia

The temperate forms of this beautiful sundew are found in Europe, eastern Canada, and the United States. The tropical forms that do not go dormant occur from central Florida south into the West Indies and

South America. The temperate forms of this species most often grow semi-aquatically around lakes in pure peat covered in 2 to 3 inches (5–7.5 cm) of water. The plants will form stems 6 inches (15 cm) high with small, spoon-shaped leaves on narrow petioles with white flowers. The whole plant is often a lovely maroon color. Tropical forms, such as those from Cuba, have smaller compact rosettes perfect for the terrarium. Very large forms are often found in New Jersey, the Carolinas, and the Gulf states. I like to grow these in undrained bowls of waterlogged peat moss, but they also do well in pots sitting in deep water and in wetter parts of bog gardens. Propagate from seed or leaf cuttings.

Temperate forms of *Drosera intermedia* are excellent in outdoor bog gardens.

Drosera × *beleziana*

The natural hybrid between *Drosera rotundifolia* and *D. intermedia* is another vigorous plant with large, upright, spoon-shaped leaves. Sterile, it must be propagated by leaf cuttings. The cultivar *D.* 'Nightmare', introduced by Ivan Snyder and discovered near the town of Toms River, New Jersey, is a particularly robust clone.

Drosera linearis

This strange sundew grows primarily in alkaline marl bogs around the Great Lakes area, often with *Sarracenia purpurea*. It has 3- to 4-inch (7.5–10 cm) linear leaves on long, upright petioles, and white flowers. As it is difficult to grow without a sustained, frozen winter, I have had limited success using a medium of one part sand to one part vermiculite, refrigerating the plants for a full six or seven months' dormancy. Folks in cold winter climates would probably find it easier to grow outdoors. The plants reportedly can adjust to acidic peat soils.

Drosera filiformis

Commonly called a "thread-leaved sundew," this very unusual and pretty sundew from North America is also controversial, for there is confusion about its status in taxonomy. Is it one species with two varieties, or two species, or two subspecies? Here I will follow the belief that it is one species with two varieties.

Thread-leaved sundews make nice additions to bog gardens, and the larger forms look beautiful in hanging pots in a greenhouse. Their dormant buds are large, blackish, and hairy. They can be propagated by seed or leaf cuttings, with the exception of the cultivar 'California Sunset', noted shortly. They also gradually form large clumps that can be divided.

Drosera filiformis var. *filiformis*

This is found in scattered locations from New England to North Carolina, with one peculiar colony, sometimes called 'Florida Giant', in northern Florida. In its typical habitat, such as the New Jersey Pine Barrens where it is most common, the plants have long, thin, threadlike leaves to 6 inches (15 cm) tall, covered in red tentacles. The flowers are pink and almost 1 inch (2.5 cm) across. The 'Florida Giant' form has leaves twice as tall. In the 1980s a colony with solid maroon leaves was discovered in Florida. These odd plants can often continue to grow through winter when kept warmer. One reason may be that the ponds around which this form grows comes from underground aquifers of warm water, so that even during frosts the leaves may die while the roots are kept much above freezing. I lost plants of this "red form" one winter when the outdoor pots froze.

Drosera filiformis var. *tracyi* growing in a water logged swamp in Florida.

Drosera filiformis var. *tracyi*

This plant, found along the Gulf states, is very similar to var. *filiformis* except the tentacles are green and the leaves reach 20 inches (50.5 cm) in length. It's a large, showy, warm temperate sundew. On a trip to the Florida Panhandle, I was thrilled to see this tenacious, large sundew growing in the wet, grassy medians right alongside the highways.

Drosera 'California Sunset'.

Drosera 'California Sunset'

This is a beautiful hybrid between the two varieties produced by Joe Mazrimas and published in 1981. Vigorous and clump forming, it resembles both ssp. *tracyi* with reddish glands and ssp. *filiformis* 'Florida Giant'. Since *Drosera* 'California Sunset' is a cultivar, it should be reproduced only through leaf cuttings, and the seed should be destroyed.

The flower of *D.* 'California Sunset'.

Drosera × *hybrida*

Rarely has a sundew caused as much controversy as this little oddity. The famous botanist John MacFarlane discovered several of the plants in the 1890s while exploring a New Jersey bog with his students. He removed the clump of eleven plants and identified them as a hybrid between *D. intermedia* and *filiformis*. No one saw this hybrid again until Rich Sivertsen and Dave Kutt rediscovered it at Lake Absegami near the New Jersey shore in 1974 and distributed it among collectors. Ironically, as a teenager I lived near that lake and had seen the plants there a couple of years before. They grew in an area only a few feet (1.8 m) wide; I thought they were a stunted form of *D. filiformis* and left them alone. In 1978 at another nearby location, Phil Sheridan, Jim Bockowski, and Mike Hunt also found this hybrid. As of 2011, no plants of this cross seem to

The elusive *Drosera* x *hybrida* from the exotic New Jersey Pine Barrens.

exist any more at these locations. John Brittnacher wrote a well-researched piece on *D.* × *hybrida* in the December 2011 issue of *CPN.*

However, in 2012 other locations had been found. In the 1970s, a famous CP grower introduced a few plants of this sterile hybrid to Butterfly Valley, a botanical preserve in California's Sierra Nevada. By the early 2000s, another well-known CP grower was removing the plants, since they were not native to this preserve, but strangely they kept reappearing. In 2011, a friend of mine, Harry Tryon, who had removed a few plants and was growing them in his collection, told me his plants were producing seed! This turned out to be true, and I brought it to the attention of the society: the normally sterile *D.* × *hybrida* had mutated in the California bog in a fashion possibly similar to the species *D. anglica* originating from *D. rotundifolia* and *linearis*! Some people feel that this is an evolutionary stepping stone toward the creation of a new species.

D. × *hybrida* produces linear leaves up to 3 inches (7.5 cm) in height and requires several months of cold winter dormancy. The typical Lake Absegami clone is sterile and redder than the fertile Butterfly Valley plant and is excellent in bog gardens.

Drosera arcturi

This rare cold temperate species grows in alpine regions of New Zealand, Tasmania, and Australia, where its mountain habitat is often snow covered in winter. The plants prefer sphagnum moss, and they die down to an unusual cone-shaped hibernaculum. The peculiar, strap-shaped leaves are 3 to 4 inches (7.5–10 cm) long and an odd purplish brown color. My own attempt with this plant failed, as my winters are too warm to sustain its dormancy. It needs a cool summer as well. This weird species is considered a relic from ancient times.

Fork-Leaved Sundews

This group of sundews is restricted to the east coast of Australia, with one species found also in Tasmania and New Zealand. Usually referred to as varieties of one species, *Drosera binata*, their taxonomy is still doubtful. It's hoped that current research will clarify the matter, and thus the names of some forms may change. Some of the names have been used for convenience and have no legitimate value.

Forked sundews are all magnificent, showy species. They grow from tropical to warm temperate climates, in wet swamps and on dripping cliffs that may experience frost in winter.

The plants have thick black roots from which long petioles arise. The leaves themselves branch into two or many points, depending on the varieties. Often the branching, tentacle-covered leaves resemble large, ferny spiderwebs. The flowers also are large, usually white but sometimes pink, in clusters on tall stems. As they rarely produce seed in cultivation (they need to be crossed with other clones and most plants in cultivation originated from single individuals), growers usually remove the flowers due to their exhausting effect on the plant.

Forked sundews have been popular plants since Victorian times. The larger forms look best in large hanging pots that are watered daily, or in pots that sit in hanging bowls of water. They also do well on the tray system, but the drooping leaves of most forms will have their beauty masked. Forked sundews also succeed quite well planted in undrained containers, such as oversized brandy snifters or tall flower vases, which give ample room for their leaves to hang luxuriously.

In suitable climates, they are unbeatable for the bog garden. They make attractive flycatchers for very sunny porches and patios. Most are too large for the average terrarium, but they can succeed if dormancy is respected and the light levels are high enough. Sometimes they do well on windowsills—but only with several hours of sun and in humid conditions. In cool and warm greenhouses, they are unsurpassed.

These *Drosera* do equally well in peat and sand/perlite mixes or long-fibered sphagnum. They are easy to propagate from root and leaf cuttings. Young plants, and the first leaves of the season on older ones, usually fork once, like *Drosera binata*, but will rapidly attain their full form in time. Because of their lengthy Latin names and general lack of common names, these plants are usually referred to in conversation by their varietal names, such as "*Drosera dichotoma*," dropping the species "*binata*."

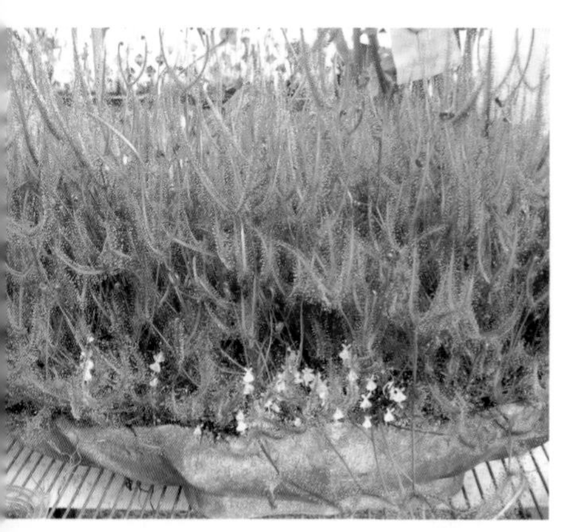

Drosera binata var. *binata*, growing in an undrained ceramic container with the flowering *Utricularia sandersonii*.

Drosera binata var. *dichotoma* 'Giant', an enormous forked sundew and one of the author's favorites.

Drosera binata var. *binata*

Considered by many to be the ancestor of all forked sundews, this species is actually uncommon in its native Australian swamps. It is also found in New Zealand. More upright and compact than its sisters, *Drosera binata* var. *binata* has leaves up to a foot (0.3 m) tall that fork once. Often called "T-form," the leaves are actually Y-shaped and are either green with intensely red tentacles or entirely red. The plant spreads into massive clumps. Dormant in winter, it is tolerant of brief freezes to 15°F (-9°C), but if the roots freeze completely you may lose the plant. The flowers are white.

Drosera binata var. *dichotoma* 'Giant'

This is one of the most massive of *Drosera*. The leaves are olive to bronzy yellow, with nearly transparent tentacles and pink glands. The wiry petioles can be a foot (0.3 m) in length, and the leaves branch from four to twelve points, sometimes 2 feet (0.6 m) in diameter. (Visitors to my nursery have involuntarily screamed when they've accidentally backed into my 4-foot-diameter [1.2 m] specimens!) A superb flycatcher, this species has a brief winter dormancy and is as cold tolerant as *Drosera binata* var. *binata*. Outdoors in full sun, the leaves are colorful and smaller and held erect. The flowers are white and best removed. The larger the pot, the more massive the plants become. This variety is also commonly called *Drosera dichotoma*, and I have registered it as the cultivar *Drosera* 'Giant'. A common name is the staghorn sundew. In New South Wales, Australia, millions of this form grow on dripping wet canyon cliffs, their huge forked leaves hanging downward.

Drosera binata var. _dichotoma_ 'T-Form'
This handsome, compact variety has the exact same yellow-green coloration as 'Giant' but is somewhat smaller, producing leaves that fork once into a "T" or "U" shape; very rarely more points are produced.

The large flowers of *Drosera binata* var. *dichotoma* 'T-Form'.

Drosera binata var. _multifida_
This large forked sundew has drooping, olive-colored leaves with red tentacles that can branch a dozen or more times, and white flowers. It usually has a dormant period in the winter.

Drosera binata var. _multifida_ 'Pink Flowers'
Identical to the above but with attractive pearly pink flowers.

Drosera binata var. _multifida_ 'Triffida'
In the 1980s I was given a forked sundew tagged with the name *D. dichotoma* 'small red form'. I have sold and circulated this plant both intentionally and unintentionally under that name. Unintentionally—because this variety of *D. binata* is the only forked sundew I have encountered that self-pollinates its white flowers and spreads

Drosera 'Triffida', a rapidly seed-spreading forked sundew aptly named after *Day of the Triffids*.

everywhere, as prolifically as cape sundews! It is quite unlike the varieties *dichotoma*, having bright red tentacles on green leaves that fork once to several times. Most striking is that mature plants flower in midsummer and then go dormant early as their numerous seeds germinate all about them. Mother plants return to growth the following year. I have registered this plant as the cultivar *Drosera* 'Triffida' after John Wyndham's nightmare monster plant novel *The Day of the Triffids*.

Drosera binata var. *multifida* form *extrema*

Perhaps the most beautiful of the forked *Drosera*, this form was found on Stradbroke Island off southern Queensland and on surrounding mainland areas. It's a lovely plant with pendulous leaves over a foot (0.3 m) long that branch enormously, forming globose webs with forty to seventy points! In strong light the whole plant is a rich maroon color. The white flowers are best removed. Keep over 40°F (4°C) to avoid dormancy. The plants will return from brief, hard freezes.

Drosera 'Marston Dragon'

Adrian Slack created this attractive, large hybrid between *Drosera binata* var. *dichotoma* 'Giant' and *Drosera binata* var. *multifida extrema*. The leaves are unusual in that the forked tendrils are hooked, reminding Slack of the taloned feet of a Chinese dragon. The leaves are dark green with red tentacles, and fork to eight or so points. This cultivar usually has a brief winter dormancy.

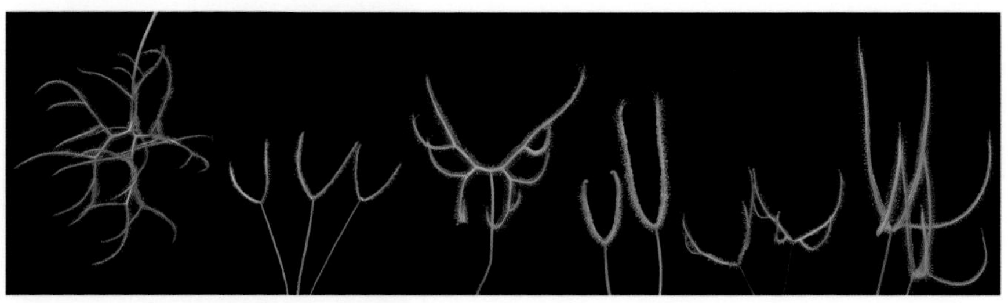

The leaves of forked sundews. Left to right: *Drosera binata* var. *multifida* f. *extrema*, *Drosera binata* var. *binata*, *D.* 'Marston Dragon', *D. binata* var. *dichotoma* 'T-Form', *D. dichotoma* 'Small Red', and *D. dichotoma* 'Giant'.

Tropical Sundews

Because of their exotic appearance, people assume that *all* sundews come from exotic, tropical places, but sundews from truly tropical places are not as common as one might think. I have already discussed some of these in the previous sections. Many forms of *Drosera spatulata* come from tropical climates, as do *Drosera burmanni*, *D. villosa*, *D. anglica* 'Hawaii', *D. multifida extrema*, and forms of *Drosera intermedia*. But even some of these sundews are also tolerant of occasional chilly temperatures or even frost, and *Drosera burmanni*, an annual, can be grown as a summer plant in temperate climates. Also keep in mind that many sundews, even when found near the equator, are often mountain plants used to temperatures much cooler than in the oppressively hot lowlands.

Here I will discuss *Drosera* that are best grown warmer year-round. Some of these, too, may tolerate chillier temperatures, and as growers experiment with them we may learn more of their tolerance extremes. But suffice it to say, the following sundews generally do best when temperatures remain above 50°F to 60°F (10–16°C), as in a terrarium under grow lights or in warm and hot greenhouses.

Three tropical species from northern Australia deserve mention. All behave as annuals, renewing themselves from seed during the wet season, but may grow longer when kept wet. Follow the cultivation techniques for the woolly sundews, described shortly.

Drosera indica

This species is also found in other tropical countries such as Africa and India. The plants look surprisingly like *Byblis liniflora* and are often mistaken for them. *Drosera indica* produce a scrambling stem 6 inches (15 cm) in length. Long, linear leaves covered with tentacles radiate out from this stem in all directions. The flowers can be white, pink, or even orange. A pretty and delicate species for the hothouse or terrarium.

Drosera banksii

This curious-looking small species appears similar to the climbing tuberous *Drosera* that will be discussed later. However, these plants produce no drought-resisting tubers and instead die away during the dry season. The plants are erect and 4 inches (10 cm) tall. They produce a few cup-shaped, peltate leaves on thin petioles. Propagate from seed. A similar species is *D. subtilis*.

Drosera hartmeyerorum.

Drosera hartmeyerorum

In 1995, while filming a video on CPs in the Kununurra area of northern Australia, German enthusiasts Siegfried and Irmgard Hartmeyer found an unusual red sundew very similar to *D. indica*. On seeing the film, Dr. Jan Schlauer believed it to be a new species. The Hartmeyers returned to Australia in 2001 and rediscovered this most unusual plant, which Schlauer named for them after studying specimens.

Closely related to *D. indica*, this new species has several bizarre tentacles at the base of the linear leaves where they join the stem; the tentacles are not sticky and appear to be yellow, light-reflecting lenses! In the dark, when a flashlight illuminates them, they "glow"! Further studies seem to prove these reflectors lure insects to alight on their sticky leaves. Nothing similar is known among sundews or anywhere else in the plant kingdom.

Three Sisters from Queensland

I call the following three tropical sundews "the three sisters" because not only are they related, but they also grow in similar conditions: humid, shaded, tropical rain forest with little temperature extremes. All come from Queensland, Australia, usually along the damp, mossy edges of slow-moving streams. Although rare in the wild, they are very popular in cultivation, not only for their unusual appearance but also because they are ideal for the potted terrarium under grow lights. They despise strong direct sunlight. They prefer being grown in long-fibered sphagnum. They are easy to propagate from leaf and root cuttings, rarely producing seed. They are happiest with high humidity and a temperature range of 65°F to 80°F (18–27°C). These three sundews also do well by a sunless, north-facing window if placed under a bell jar or in a small tank. Warning: Most fertilizers should never be applied to these plants; they will fry the leaves. Only Maxsea is safe.

Drosera schizandra

Commonly called the notched sundew, this odd plant produces a flat rosette of broad, nearly oval leaves. But as it grows older the leaves may elongate and develop a notch at the end, achieving a heart-shaped form. The tentacles are sparse and weak, catching mostly gnat-size insects. The beautiful and delicate rosettes are typically 4 to 6 inches (10–15 cm) in diameter. The plants commonly form clumps that develop from their roots. A short scape produces unusual small, red-petaled flowers.

Three sisters from Queensland. Clockwise from upper left: *Drosera schizandra, D. prolifera, D. adelae.*

Drosera prolifera

The "hen and chicks" sundew has oval to kidney-shaped leaves up to 3/4 inch (2 cm) across, held semierect on long, thin petioles. Most peculiar are the flower stalks, which run along the ground bearing a few small red flowers, ending in a small baby plant if the tip touches the ground. Over time, with frequent flowering, many new plants take root around the "mother," and sometimes the flowers produce seed. The dime-size leaves will fold around prey like a sandwich.

The flowers of *Drosera adelae* are like miniature red stars.

Drosera adelae

The lance-leaved sundew is the errant sister, for she can take somewhat brighter light, less humidity, and cooler temperatures. A lovely sundew for the terrarium, she can also occasionally thrive on bright, humid windowsills (for example, a bright bathroom). The leaves are arching, long, and pointed. In lower light, the plants are green and can reach a foot (0.3 m) across. In brighter conditions she will develop stiffer, shorter, bronzy leaves. The stunning flowers look like bright red stars, and many dozens can appear on stalks 6 inches (15 cm) tall. *Drosera adelae* is a spreader: older plants may die away and be replaced by many offspring that come up from the roots. A curious fact is that originally only plants with dull, yellow-white flowers were known. After the introduction of the red-flowered form (which some believed was a new species because the flower structure also seemed different), the pale-flowered form seemed to disappear—or turn red. On several occasions I have been given fresh material of the older variety, but these in time also turn red!

Cooler Growing Tropical Sundews

While these sundews grow in the tropical areas of the world, they are often mountain plants, enjoying cooler temperatures, especially at night.

Drosera madagascariensis, from Botswana, Africa, given to the author by *Power Rangers* actor Ed Neil.

Drosera madagascariensis

In the early 1990s an actor on the *Power Rangers* television and movie series, Ed Neil, brought me a wet bag of some sundew leaves he had found growing in a ditch near the Botswana airport in Africa a few days before. I grew plants from the leaf cuttings and sold them as *D.* 'Botswana' before correctly identifying them as this species, which grows on the island of Madagascar as well as in parts of continental Africa. An attractive species for the terrarium, it grows stems 4 to 8 inches (10–20 cm) long topped with short spoon-shaped leaves, and new shoots will appear from the base of the plant as older stems die off. In warm conditions the plants grow continuously, but they also return after chilly winters, and the tall scapes of pink flowers self-pollinate, producing abundant seed.

Drosera ultramafica

Recently discovered in the mountains of Palawan in the Philippines, this attractive sundew has numerous elongated, strap-shaped leaves.

Drosera chrysolepis, a highland tropical sundew.

Drosera chrysolepis

This species from Brazil is usually found growing in quartz sand and rubble. The handsome leaves are held upright on a short stem and have an elongated diamond shape, tapering to a point.

Drosera roraima and *D. solaris*

These two similar sundews come from the high tepui plateau of Mt. Roraima and the surrounding Venezuelan Guiana Highlands, and slowly produce erect stems to 8 inches (20 cm) high cloaked in dead leaves like thatched skirts. The rosettes of living leaves are about 2 inches (5 cm) across and are spoon-shaped in *D. roraima* and oval in *D. solaris*. Since these are highland tropical sundews, they do best with cooler temperatures from 40°F to 70°F (4–21°C).

Drosera graminifolia var. *spiralis*, easy to grow but problematic to propagate. Root cuttings may work.

Drosera graminifolia

Reminiscent of *D. filiformis* from North America, but with fewer and thicker linear leaves to 8 inches (20 cm) long, this Brazilian species is also a highland tropical. It grew successfully for me for several years, but I could never multiply it from leaf cuttings, and the pink flowers never produced seed.

Drosera hilaris
From South Africa, this rare species looks similar to *D. capensis* but has wide, paddle-shaped leaves and will go dormant during periods of frost, reemerging from the roots or stem.

Woolly Sundews

An exciting development in the past few decades has been the discovery of many new sundews in the vast, remote areas of northern Australia. For a long time one species, *Drosera petiolaris*, the woolly sundew, was the only type formally described. But now, thanks to the enthusiastic work of people like Allen Lowrie and Katsuhiko Kondo, a whole group of related *Drosera* known as the *petiolaris* complex is beginning to enter cultivation.

Many of these plants are still mysterious and rare and are only now beginning to be identified and named. Some of them currently are known by location names only, but this is rapidly changing as the plants are formally described. Cultivation techniques are also being explored.

Many of these sundews are notable for their unusual petioles and their leaf blades densely packed with long, fine tentacles.

Drosera petiolaris x *dilatato-petiolaris*, with *D. burmanni* in the foreground.

Northern Australia has a warm tropical climate marked by six months of rain and six months of drought. These sundews grow in sandy areas that are wet to waterlogged during the rainy season. When the soils dry out during the annual drought, most of these plants die down to their thick, often hairy, bulblike stems for several months of dormancy. However, in cultivation the plants remain in growth if they are kept wet.

To grow these types, I recommend a soil recipe of two parts sand to one part peat. Use the tray system in brightly lit heated terrariums or a sunny hothouse. Temperatures are best between 60°F and 90°F (16–32°C). Propagation is by leaf cuttings, division, and sometimes seed. They can be grown outdoors in tropical climates.

All of the following sundews are beautiful and unique, and vigorous hybrids are common. Here are some of the more popular forms.

Drosera petiolaris

The original species, this plant was discovered by Joseph Banks in Queensland during Cook's voyage around 1770. Fond of waterlogged conditions during the wet season, the rosettes are 3 to 4 inches (7.5–10 cm) across with many erect leaves that have long, narrow petioles and small, spade-shaped traps with very long retentive glands. The flowers are dark pink.

Drosera dilatato-petiolaris

Although this species was first recognized by L. Diels in 1906, Kondo formally described it in the 1980s. Common in damp areas around Darwin (on the coast of the Northern Territory), this clump-producing species has medium-broad green petioles and small, circular red traps, most of which lie flat along the ground. Clumps can be over a foot (0.3 m) across. The flowers are white.

Drosera falconeri

Discovered around 1980, this desirable species has short, broad petioles with immense, oblong traps over 1 inch (2.5 cm) wide. The leaves are pressed flat to the ground and the whole plant is a deep maroon color. It is now known to be common in the coastal areas of the Northern Territory. Since the oval-shaped leaves often fold like a taco shell when eating larger prey, this type of sundew is sometimes thought to be a precursor of the Venus flytrap in evolution.

Drosera falconeri.

Drosera lanata

A pretty species with clusters of low-growing leaves that have long thin petioles and small circular leaf blades. The leaves and center of the rosette are covered with dense, silvery, woolly hairs.

Drosera ordensis, a discovery of Allen Lowrie in the desolate Kununurra area of northern Australia.

Drosera ordensis

This stunning sundew was discovered by Allen Lowrie and was called *petiolaris* 'Kununurra'. The rosettes are 3 to 6 inches (7.5–15 cm) in diameter and can form clumps over a foot (0.3 m) across. The leaves are erect, with long, wide petioles so densely covered in silvery hairs that they appear white. The small circular traps are golden green. The large flowers can be pink or white.

Drosera paradoxa

This unusual form is very similar in appearance to *Drosera petiolaris* except the leaves are hairy, like those of *Drosera lanata*. In the sun the plants grow in pincushionlike rosettes on the ground. When shaded by bushes, they can grow to a foot (0.3 m) tall, leaning among other plants.

Pygmy Sundews

Pygmy sundews, almost all of which originate in southwestern Western Australia, are a fascinating and complex group of plants. Most of these plants are true miniatures, rarely larger than a penny, although a few can be somewhat larger in size. All are beautiful, diminutive jewels, best seen in colonies with the aid of magnifying glasses. There are probably around forty or so species, including a few hybrids.

The climate of Western Australia is considered warm temperate to subtropical and Mediterranean. This means the summers are very hot and very dry. Winters are cool, with night temperatures in the 30s°F and 40s°F (-2–4°C) and day temperatures between 50°F and 65°F (10–18°C). Most of the rain falls during the winter months, and it is then that the flora of the region does its most vigorous growing. This is also true of the carnivores from this region of the world. As the hot days of summer approach, most of the plant life goes dormant while desertlike conditions prevail. The soil is predominantly sand.

Pygmy sundews are typically small rosetted plants rarely over an inch (2.5 cm) in diameter. The leaves radiate outward on short petioles with small, tentacle-covered traps that are circular to spoon-shaped. The tentacles can be very long around the edge of the leaf, and move quickly when small prey are trapped. The center of the plant usually has a cone-shaped structure of shiny, dense hairs called stipules. In summer, when pygmies are dormant, the plants fold up to these heat-reflecting tufts of hair.

Now you know why they're called pygmy sundews. This one is *Drosera nitidula* x *pulchella*.

A most curious aspect of pygmy sundews is their ability to produce gemmae, or brood bodies—an asexual method of reproduction. Each gemma is pinhead size or smaller and may resemble the scales of a fish or small beads.

Clusters of gemmae arise out of the stipules, forming crowns of these brood bodies in the center of the plant. This occurs in autumn as the days get cooler and shorter and the winter rains begin. Each gemma is held by a tensely coiled hair, and when struck by a raindrop, the gemmae explode outward—sometimes shooting many feet or meters from the mother plant! Each gemma quickly takes root and rapidly grows into a new plant, genetically identical to its parent. The production of gemmae makes pygmy sundews easy plants to propagate.

Gemmae brood bodies arising from the center of *Drosera nitidula*.

The flowers of the pygmies are also interesting. While some flowers are small, many are large, showy, and in a wide range of colors. Some flowers are larger than the plants themselves. However, in cultivation they rarely produce seed, so most growers rely on the dependable production of gemmae for propagation. To the delight of growers, many of the flowers are also highly fragrant—a pleasant bonus for such tiny plants.

The typical life cycle for pygmies in the wild is as follows. After their summer dormancy, as the rains begin in autumn, a few leaves are produced, followed by gemmae. By early winter the gemmae are scattered and the plants grow vigorously in their wet, sandy soil. Winter and spring are their most active growing season. By late spring the plants flower and are cross-pollinated by insects. By early summer, seed is set, the soil dries out, and the plants go dormant. Seed germinates the following wet season. In some low-lying areas, pygmies may be flooded and killed in springtime after heavy rains. But a few plants always survive on higher ground, and the following winter they rapidly replenish their colonies through massive gemmae production, only to be killed off again the following spring.

D. paleacea ssp. *paleacea*

Left to right: The substantial flowers of *Drosera callistos* and *D. hyperostigma.*

Most pygmy sundews are easy to grow in cultivation. Fortunately, when kept wet year-round, most species don't require a dry summer dormancy. Therefore, they can be enjoyed year-round and more or less grown like typical subtropical rosetted sundews.

Pygmy *Drosera* grow well in a half-sand to half-peat mixture. They also do well in about 80 percent peat to 20 percent perlite. Grow them in the tray system unless I indicate otherwise in the following sections. A sunny position is required. All thrive in cool and warm greenhouses. They also do well in terrariums grown close to the lights, but you should shorten the photoperiod in autumn for gemmae production. They are superb outdoors in pots and bog gardens in warm temperate, subtropical, and Mediterranean climates and will spread in bogs to form beautiful glittering colonies. They are generally tolerant of brief, light frosts into the mid 20s°F (-4°C) but perform even better in frost-free climates. On windowsills they require very sunny conditions.

Although pygmy *Drosera* primarily catch small, gnat-size insects, very large prey such as crane flies, moths, and houseflies are also caught. Sometimes dense colonies of the plants act in a group effort to overpower prey. Struggling insects leave behind their legs and wings in their attempt to escape the long, rapidly moving tentacles. The individual plants feast on whatever insect parts they can grab.

The following are some of the most popular pygmy sundew varieties.

Drosera callistos* and *D. hyperostigma

These handsome pygmies are up to 1 inch (2.5 cm) in diameter with elliptic leaf blades and large, shiny, orange flowers. If these species go dormant in hot summers, keep the soil just barely damp.

Drosera closterostigma

Bright red rosettes up to 2/3 inch (1.7 cm) in diameter. The large flowers are white with red centers.

Drosera dichrosepala* and *D. enodes

These species have 1/2-inch (1.3 cm) rosettes of spoon-shaped leaves clustered atop a slow-growing stem. The white flowers are sweetly perfumed. Keep them almost dry if they go dormant.

Drosera eneabba

Bright red rosettes up to 1 inch (2.5 cm) in diameter, with pretty pinkish white flowers with scalloped edges.

Golden green rosettes of *Drosera enodes*.

Drosera leucoblasta
Small reddish rosettes of circular leaves less than 1 inch (2.5 cm) in diameter. Bright orange flowers are as large as the plant. This species may go dormant in summer.

Drosera mannii
A beautiful, vigorous pygmy with broad petioles and elliptic red traps. The large flowers are a pearly pink or white.

Drosera nitidula
This species has attractive reddish leaves with white flowers and is known to hybridize with other pygmies.

Drosera occidentalis ssp. *occidentalis*
One of the tiniest, and a favorite of mine (under a powerful magnifier!). Minute, fine, sparse, deep red leaves, with pink flowers.

Drosera paleacea ssp. *paleacea*
Rosettes 1/2 inch (1.3 cm) across with many small, white, perfumed flowers that smell like coconut.

Drosera pulchella, a robust pygmy sundew with variably colored flowers.

Drosera pulchella
Robust rosettes up to 1 inch (2.5 cm) in diameter, with flowers ranging from white to pink to orange. The broad petioles and circular reddish orange leaves are very attractive.

Drosera pygmaea
The only pygmy found outside Western Australia, this tiny species is also found in southeastern Australia, Tasmania, and New Zealand. Fascinating for its small, bright red to green rosettes and minute, four-petaled white flowers.

Drosera citrina
This species has stunning white flowers with yellow borders.

Drosera scorpiodes

This is a "giant" pygmy, exceptional for its large rosettes that measure over an inch (2.5 cm) across and its tall, rootlike stems. The dished, lance-shaped leaves have exceptionally long tentacles. I grow two forms. The larger form, with white flowers, grows rapidly but is short-lived, dying away after two or three seasons. By contrast, my plants from the smaller form with dark pink flowers have grown for more than seven years, and have stems 6 to 7 inches (15–18 cm) tall.

Drosera scorpiodes, a giant among pygmies.

Drosera lasiantha

Impressive flowers on a hairy stem make this species eye-catching.

The flowers of some pygmies can be as big as the plant! This is *Drosera lasiantha*.

Drosera patens × *occidentalis* ssp. *occidentalis*

Bright reddish rosettes up to 1/2 inch (1.3 cm) in diameter. The flowers are white with bright red stigmas. Now believed to be extinct in its original range due to habitat destruction. Ironically, this tiny plant survives by the thousands after its introduction into Mendocino County, California. A vigorous hybrid, it is probably the most popular in cultivation along with *Drosera nitidula* × *pulchella*, below.

A dense colony of *Drosera patens* x *occidentalis* ssp. *occidentalis*, extinct in Australia but introduced to California.

Drosera nitidula × *pulchella*

My favorite of all the pygmies. The deeply colored rosettes are very robust, up to 1 inch (2.5 cm) across, and similar to its parent, *Drosera pulchella*. It grows as easily as any cape sundew, and the lovely flowers are a showy, pearly pink.

Drosera meristocaulis

An incredible 1950s discovery was this species found growing on the tabletop tepui Mt. Neblina on the border of Venezuela and Brazil. It is considered an ancient surviving relic, much like the discovery of *D. sessilifolia* in the same area, which is nearly identical to the Australasian species *D. burmanni*. *D. meristocaulis* was recently rediscovered and has been shown through DNA research to be closely related to the pygmy sundews on the other side of the world. Tiny rosettes of red spoon-shaped leaves top scrambling stems that take decades to form, and the lovely pink flowers grow not on tall stalks but emerging from the crown of leaves. It does not, however, produce gemmae or seed, being self-sterile, but has been tissue cultured.

Tuberous Sundews

Western Australia is home to another large group of sundews that share the same wet winter–dry summer habitat as the pygmies. These are the tuberous *Drosera*. Instead of dying down to protective cones of stipule hairs to survive the dry heat of summer, these plants have adapted to a dormancy underground, where they hibernate as small tubers.

The tuberous *Drosera* are about as strange and as beautiful as sundews get. The fifty or more species come in a wide range of growth habits. Erect sundews have self-supporting stems with small, tentacle-covered leaves scattered along their length, sometimes reminiscent of little trees. Climbing sundews grasp and loop over bushes and other plants as though they were vines. Fan-leaved sundews can be so elaborate in their design that they can appear like several different sundew species all glued together. Tuberous sundews can also be rosetted, often covering the sandy soils with dense mats of carnivorous leaves—a lethal landing platform from which few insects can escape.

The tubers of these plants may be pea- to walnut-size. They can extend underground at a depth of a few inches (7.5–15 cm) to a couple of feet (0.3–0.6 m). The tubers can be brightly colored in oranges and reds. They are often covered with insulating papery sheaths, each layer marking a year's growth. It is in the tuber that the plant stores its energy reserves.

On the top of the tuber is a small "eye," not unlike that of a potato. As autumn approaches, a vertical stolon grows from this eye to the soil surface, even when the soil is still dry. Usually the plant's roots form along this stolon. Sometimes the root mass is so thick it forms a "chimney" through which the tuber grows and recedes each season. Energy from the tuber gives the plant its resources to develop above ground during the cool, wet winter. For five or six months the tuberous sundews catch insects and flower and do all the things happy carnivorous plants do. Then summer approaches, and the days get longer, hotter, and drier. The plants turn brown and die back, and all the minerals and energy of the plant return down the withering stolon into the old tuber. Usually the tubers get larger each year. Often the plants produce more food than one tuber can store. In this case, new stolons develop horizontally from the mother tuber, and new tubers are grown. Some species don't return to their old tuber but create a new one each season adjacent to the dried shell of the older one.

Drosera hookeri with prey. The tentacles pull the victim into the cup-shaped leaves.

Tuberous sundews are still uncommon plants in cultivation, but they are growing in popularity as more plants are propagated and as growers become less intimidated by the plants' undeserved reputation for being difficult. I have found this group to be challenging but fairly easy to grow. The most problematic challenge is obtaining plant material. Currently, most plants in cultivation are grown from seed or tubers imported from Australia. Propagation can be difficult because most of the species don't produce seed in cultivation and are difficult to grow from leaf cuttings. Production of additional tubers is a common but slow affair.

If you can obtain a potted, established tuberous sundew, consider yourself lucky. If you receive tubers imported from Australia and you live in the northern hemisphere, remember that the seasons are reversed, so

Drosera hookeri.

most likely you will receive your dormant plants during Australia's summer (December to March), which is the northern hemisphere's winter growing season for these plants. Therefore, due to the plant's biological clock, the tuber begins to grow as summer starts. This first crucial year of seasonal adaptation is the most challenging aspect of establishing the plants.

When you receive your tubers, store them in airtight plastic bags with a few strands of barely damp sphagnum moss and keep them at room temperature in a dark place such as a drawer. Check them weekly until you notice the stolon's growth beginning from the eye of the tuber.

Pot them using 5- to 8-inch (12.5–20 cm) drained plastic containers in a mix of prewetted two parts sand to one part peat. With your finger, make a hole in the center of the soil about 2 to 4 inches (5–10 cm) deep. (Some large tubers may need a bigger pot and a deeper hole.) Using forceps, lower the tuber into the hole so the eye is facing upward. (Remember, the eye is where the stolon is emerging.) Then fill the hole with pure sand, but don't pack it tightly. Set the pot on the tray system, maintaining the water level around 1 inch (2.5 cm).

Within a few weeks, the plant should emerge and develop quickly into its full size. Most tuberous *Drosera* do not continually grow new leaves all season long. The plants produce all their leaf growth rapidly and retain those leaves for the duration of the season.

The objective is to sustain the plant during a shortened summer growing season. Keep the plants as cool as possible—for example, near swamp coolers in the greenhouse or in a cool terrarium under grow lights. During this first season of adjustment they usually grow for around three months, from around April or June until July or August. Feed the plants insects during this time. When the plants turn brown and die back, remove the pot from the water tray. Set the pot in a shady place in moderate temperature—for example, under a greenhouse bench. The soil must now dry out completely.

Usually the plants remain dormant through autumn and return to growth in wintertime. Do not set the pot back in the water tray until you see the stolon emerge from the soil. Sprinkle water from overhead to thoroughly wet the soil. Allow the plants to grow normally until another cycle of dormancy sets in.

When adjusted to the northern hemisphere, tuberous sundews begin to grow (usually) between October and December. They go dormant between April and June. The plants enjoy sunny conditions in cool and warm greenhouses during the winter months, in temperatures best kept on the cool side: 40°F to 50°F (4–10°C) at minimum, 60°F to 75°F (16–24°C) maximum. I think they would thrive in the cooler, potted terrarium on a winter photoperiod, though I have not yet tried this. Remove the pots during summer dormancy and dry them out. Tuberous sundews make nice winter replacement plants if you remove dormant temperate species from your tank, such as Venus flytraps or winter-dormant sundews.

Some growers prefer to remove the dormant tubers from their pots every summer, storing them dry in airtight plastic bags. I prefer to leave them potted, but I transplant them every three years or so. This is usually necessary to separate additional tubers that have been produced.

You may be able to grow tuberous *Drosera* as outdoor potted plants if you live in a frost-free Mediterranean or subtropical climate such as along the California coast or central Florida. Protect the plants if temperatures drop below freezing. Never leave pots of dried, dormant plants in direct sunlight, as the heat will desiccate the tuber. You might also try these plants on a cool, sunny, humid, south-facing windowsill in winter.

The flowers of tuberous sundews are usually very pretty, often large, and very sweetly fragrant. Some species flower early in their growth cycle, others late—just prior to dormancy. Nearly all are self-sterile and need to be crossed with other genetic clones of the same species to produce viable seed.

Erect Tuberous Sundews

These sundews produce stems that are usually self-supporting but may occasionally lean on other plants. The leaf traps are often small, barely 1/4 inch (6 mm) across, yet they are often strong enough to catch a housefly of equal size.

Drosera peltata complex

For two centuries plants in this related group were known as *Drosera peltata* or *D. auriculata*, but in 2011 Robert Gibson argued, in a detailed paper, that the group needs to be broken into several species. This complex includes *D. lunata*, the only tuberous sundew known to exist outside Western Australia, growing also on the east coast of that country and in New Zealand, India, and much of Southeast Asia. The most common form grown by myself and others, *D. peltata*, has been identified by Gibson as *D. hookeri,* and the following comments pertain mostly to that species.

D. hookeri produces a basal rosette of nearly oval leaves early in the season, not unlike a normal rosetted sundew. Then the climbing stem emerges, carrying several shield-shaped, peltate leaves along its length. The stems average 6 to 10 inches (15–25 cm) tall. At the top of this stem, the plants have a few small white flowers. Often the stem continues to grow after blooming. This species self-pollinates, producing abundant seed. I prefer to dry out the pots during dormancy, but I have also grown them in bog gardens in my greenhouse. In Australia, plants from this complex often come up in winter-wet lawns around homes in the suburbs, often succumbing to the lawn mower!

Drosera hookeri.

Drosera gigantea

This is the king of tuberous sundews, and one of the largest of all *Drosera*. Fortunately it is also simple to grow, and it is my favorite in this group. The red tubers can approach 1 1/2 inches (3.8 cm) across, and larger plants are best grown in 5-gallon (18 L) pots. This incredible sundew grows in the form of a small tree, with lateral branches from the stem holding many small, shieldlike leaves. The clustered white flowers come early, and the foliage is golden green to bronze. Mature plants are nearly 3 feet (0.9 m) in height.

Drosera gigantea, an easy tuberous sundew that grows like a small tree. This specimen in nearly 3 feet (0.9 m) tall.

Drosera andersoniana
A small, lovely species that forms a basal rosette similar to that of *Drosera peltata*. The stem reaches up to 10 inches (25 cm) tall, with circular, peltate leaves. The whole plant can achieve a rich red color in good light, with white to pinkish flowers topping the stem.

Drosera huegelli
A charming plant, usually under a foot (0.3 m) tall, with a few cup-shaped leaves hanging bell-like on the upper half of the stem. The flowers are large and white.

Drosera marchantii
There are two subspecies of this plant, one with pink flowers (ssp. *marchantii*), another with white (ssp. *prophylla*). The stems are rather stiff, with circular, peltate leaves.

Drosera menziesii ssp. *menziesii*
A handsome plant with an undulating stem, circular leaves, and pink flowers. This often reddish variety can approach a foot (0.3 m) in height.

Drosera microphylla
A typical erect sundew with circular, peltate leaves, but the startling flowers have large golden sepals and red petals.

Drosera zigzagia
A small, delicate reddish plant with an erect stem that curiously zigzags and lovely yellow flowers; easy to grow.

Climbing Tuberous Sundews

These plants scramble over surrounding vegetation with wiry, flexible stems sometimes many feet in length. Some of their tiny leaves cement themselves to other plants with their tentacles; these leaves usually have longer petioles than the others.

Drosera macrantha ssp. *macrantha* can scramble and climb for many feet (1.2–1.5 m). Some of its leaves have cemented themselves to the branch for support.

Drosera macrantha

Another of my favorites, this species is easy to grow. Its seed germinates easily, although cultivated plants need to be cross-pollinated with different genetic clones to produce seed. Ultimately reaching 4 or 5 feet (1.2–1.5 m) in length, the plant has small cup-shaped leaves, and the 1-inch (2.5 cm) flowers are white or pink. Seedlings remain in a small rosette the first year of growth, forming short stems by the second year. One specimen of mine has remained in the same gallon-size (3.8 L) pot for more than seven years and puts on quite a show each winter and spring.

Drosera modesta

This greenish climber, almost 3 feet (0.9 m) tall, has shield-shaped leaves with extra-long tentacles and white flowers.

Drosera subhirtella

Rather fine, reddish plants with circular peltate leaves and lovely yellow flowers.

Fan-Leaved Tuberous Sundews

These are some of the most highly developed *Drosera* in the world, and some are quite robust and unusual. Most produce basal rosettes of rather flat oval to elliptical leaves, from which one to several stems grow, often with whorls of folded, fan-shaped leaves along its length.

Drosera stolonifera complex

Taxonomical lumpers and splitters have argued over this species and its cousins for over a century, and in recent years all subspecies have been elevated to species status. *D. stolonifera* is my favorite, a thick-stemmed, robust species with whorls of reniform, or kidney-shaped leaves that are somewhat cupped or folded. Several stems may appear, each around 6 inches (15 cm) long, with leaves in several groups of three to five scattered along the stem's length. The large flowers are white. *D. purpurascens* (once ssp. *compacta*) is rather smaller, deep red, with compact stems and nearly funnel-shaped leaves. *Drosera humilis* (formerly ssp. *humilis*) is a delicate, reddish plant whose stems are semierect, the small leaves on thin petioles. *Drosera prostrata* (originally ssp. *prostrata*) has interesting stems that trail along the ground, and whorls of nearly oval leaves. Another handsome variety is *Drosera rupicola* (formerly ssp. *rupicola*), with coarse stems and large leaves that fold over prey like a sandwich.

Drosera purpurascens.

Drosera ramellosa

This small golden-green plant is like a miniature version of *Drosera stolonifera*. From a leafy rosette arise two 4-inch (10 cm) stems of folded, fan-shaped leaves.

Drosera platypoda

A fan-leaved species that I love, the plants are erect, up to 8 inches (20 cm) tall, and golden green with red folded leaves held close to the stem. The flowers are white.

Drosera rupicola in flower.

Rosetted Tuberous Sundews

Like many of their cousins around the world, many tuberous sundews grow as rosettes of flat leaves pressed to the ground. The differences are that the leaves of rosetted tuberous *Drosera* are all produced early in the season and they are often large and nearly circular, often with short petioles.

The leaves of most rosetted tuberous sundews are incapable of movement. Fragrant flowers usually appear after the leaves are formed. As a rule, the blossoms are white, and may appear clustered or as multiple single blooms.

Drosera macrophylla in flower, grown in a tub of tuberous sundews by Damon Collingsworth, co-owner of California Carnivores.

Drosera macrophylla

An easy and handsome plant for cultivation; its large leaves are teardrop shaped and the rosettes are 4 to 6 inches (10–15 cm) across. Multiple, sweetly perfumed flowers appear before the rosette is fully developed.

Drosera rosulata

Neat rosettes of obovate leaves around 2 inches (5 cm) across make this an attractive sundew, particularly when the single, short-stemmed flowers are open in the rosette's center. The leaves have a pronounced, depressed, reddish midrib.

Drosera tubaestylus

This species is also easy to cultivate. The rosettes are barely 1½ inches (3.8 cm) across, and the teardrop-shaped leaves are a bronzy red. Interestingly, this plant sometimes sends out stolons from the center of the leaves toward the end of its growing season. These stolons anchor into the ground around the mother plant prior to dormancy, where they produce tubers that develop into new plants the following season.

Drosera zonaria

An interesting and beautiful sundew with small rosettes of twenty to thirty closely held leaves arranged like overlapping shingles. The margins of the leaves are usually reddish. The plants commonly form underground stolons and develop into compact colonies.

Drosera zonaria.

Drosera orbiculata

A strange tuberous species; the leaves are almost circular at the end of long red petioles. Each rosette is about 2 inches (5 cm) across, with four to six leaves. This species was discovered in 1980.

Drosera orbiculata. Note the sandy soil Damon Collingsworth uses to cultivate winter-growing sundews.

Drosera lowriei

Another new species named after Allen Lowrie, who has done so much to widen our understanding of Australian CPs. Similar in leaf arrangement to *Drosera zonaria*, the leaves are reddish and spoon shaped and are reduced in size toward the center of the rosette.

Drosera bulbosa

There are two subspecies of this pretty sundew. *D. bulbosa* ssp. *bulbosa* looks rather similar to *Drosera rosulata* but with raised midribs on golden leaves. *D. bulbosa* ssp. *major* reaches 5 inches (12.5 cm) across and has uncommon variants with pink flowers and maroon-tinted leaves.

Drosera erythrorhiza

There are several subspecies of this plant, all producing five to twelve oval to obovate leaves. *D. erythrorhiza* ssp. *erythrorhiza* has 2-inch (5 cm) rosettes of broadly oval leaves that often form dense colonies from underground stolons. *D. erythrorhiza* ssp. *magna* has large rosettes up to 5 inches (12.5 cm) across;

Drosera lowriei.

the wide oval leaves turn reddish with age. It rarely produces dense colonies.

Drosera erythrorhiza ssp. *squamosa*, a rare beauty with leaves bordered in red.

Drosera erythrorhiza* ssp. *squamosa
One of the most beautiful and rare. The 2-inch (5 cm) rosettes have leaves with a distinctive red band along their margins.

The incredible flowers of the very rare *Drosera browniana*.

Drosera browniana
First found by orchid specialist Andrew Brown and later rediscovered by Allen Lowrie in 1992, this rare plant is easy to grow and notable for its beautiful flowers.

Drosera whittakeri
This popular tuberous sundew grows not in Western Australia but farther east in Victoria and South Australia. It is odd, because it not only grows in woodland forest habitats but also has a summer dormancy even though the soil remains damp. The 2-inch (5 cm) rosettes have oval leaves on wide petioles, with large, fragrant white flowers. Grow as you do other tuberous *Drosera*, but keep soil slightly damp during its dormancy. *D. praefolia* is nearly identical, but the flowers emerge before the leaves.

Once believed to be a variety of *Drosera whittakeri*, *D. praefolia* grows far from southwestern Australia near Adelaide and on Kangaroo Island. Flowers on left, leaf rosette on right.

South African Winter-Growing Sundews

Although most South African sundews grow year-round in permanently wet areas, some of the most unusual species from this Mediterranean climate go dormant when soils dry out, usually from midsummer to mid-autumn. This dormancy period is rather similar to that of the tuberous Australian species I have just reviewed; however, instead of tubers, these plants die down to thick, wiry roots from which they return the following season. All are easy to grow. I use a soil medium of half sand to half peat, and otherwise grow them similarly to their tuberous cousins.

Drosera cistiflora

This variable species is particularly unusual and famous for its enormous flowers. I have grown a rare purple-flowered form for decades. When growth begins, the plant forms a 2- to 3-inch (5–7.5 cm) rosette of narrow, tapered leaves pressed flat on the soil surface. After a couple of months, a stem 8 to 12 inches (20–30.5 cm) tall is formed, with narrow, tapered, strap-shaped leaves along its length. When the plants are several years old, they may flower. The sensational blooms are large, cup shaped, and 2 to 3 inches (5–7.5 cm) across. I originally believed the plants died after flowering, but this is not true. Other varieties have white, rose, or red flowers. The latter is sensational when in flower, and it emerges as a stem with few basal leaves. They rarely produce seed but are easily propagated using the earliest leaves as cuttings.

Rare in the wild, the purple-flowered form of *Drosera cistiflora* has blooms 2 to 3 inches (5–7.5 cm) across.

Drosera pauciflora

This is another impressive species closely related to *Drosera cistiflora*. It produces large, pale green rosettes of broad, wedge-shaped leaves up to 4 inches (10 cm) across. The translucent tentacles are among the fastest moving in the genus. Sometimes the rosettes creep along the ground on wiry stems, but

Drosera cistiflora, the sensational red-flowered form.

Drosera pauciflora.

they never climb. The very rarely produced flowers are equally as impressive as those of *Drosera cistiflora*. I have grown this plant for a dozen years, and just once (the first season I grew them!) I was treated to a mass flowering that lasted many months. It was a sight I will never forget, but strangely, the plants have never repeated this performance. The flower colors are as variable as those of *Drosera cistiflora*.

Drosera trinervia
A small species with flat rosettes of narrow, wedge-shaped leaves. The small flowers are white.

The King Sundew

Drosera regia, the king sundew, is in a class of its own. In nature this impressive plant is very rare—only two small colonies in South Africa are known. It is believed by some to be distantly related to the Venus flytrap, and it is considered an archaic species, one of the oldest to survive (barely!) into modern times.

The plant is large, with stiff, sword-shaped leaves that arch outward in a rosette pattern. The leaves can approach 2 feet (0.6 m) in length, and the plants gradually form stems and can develop offshoots from their thick roots. On younger plants the leaves are capable of moving dramatically, twisting in knots around large prey. On older plants only the tapered ends of the leaves move. The tentacles are substantial and produce thick globules of mucilage that can overcome even large and powerful insects. The flowers are deep pink, 1½ inches (3.8 cm) across, and held in a cluster on a stalk as long as the leaves. They can have an exhausting effect on the plant, so I usually remove them.

Use a soil mix of 80 percent peat to 20 percent perlite or sand, or even better, a high-quality long-fibered sphagnum. The king sundew grows best in cool, frost-free climates or in cool and warm greenhouses. Terrariums need to be large to accommodate it. I have seen surprisingly beautiful specimens growing on cool, sunny windowsills near the coast in California, where they appear to have a voracious appetite for houseflies. Cooler summer nights are definitely preferred.

In winter *Drosera regia* greatly reduces the size of its leaves, and occasionally crowns die back, but the plant returns from its thick roots. Leave large plants undisturbed for several years, and grow them permanently on the tray system. Leaf cuttings fail in this species, but it takes well from root cuttings. Rather than disturb the whole plant, which can set it back, I prefer to cut roots from the base of a potted specimen and gently pull these from the soil. Plants have also been propagated in tissue culture. Self-pollinating the flowers can occasionally produce seed, but the seedlings are rather slow growing.

The king sundew, *Drosera regia*.

PROPAGATION

Drosera can be propagated through various methods; most of them are easy and fun, but others can be rather challenging.

SEED: With those species that readily produce seed, this method is a rather simple affair. Some, such as cape sundews, will self-seed in such quantities that the plants can become troublesome weeds. Others may need a little coaxing. Many sundew flowers self-pollinate on closing, but with those from which you want good seed set, it is wise to tease the flowers with a toothpick while they are open. The seedpods mature in several weeks.

I prefer to separate the seed from their pods and store them in small, airtight plastic bags (jewelers have supplies of these) or paper envelopes. Seeds lose their viability at room temperature, so refrigerate them.

continued

Sowing times are simple. Almost all sundew seed is best sowed in late winter or early spring. The exceptions are tuberous and other winter-growing species, which should be sowed in early summer on dry soil and allowed to experience a couple of months of warm temperatures before wetting them in cooler autumn.

Most species germinate best with a period of stratification. This means that after sowing, the seed should experience several weeks of damp and chilly conditions. Light frost is often beneficial, even for the winter-growing varieties. However, after germination the seedlings must be protected from frost. If damping-off occurs, use a fungicide.

Sow the seed sparsely on the medium the plants prefer. Do not cover the seed with soil. High humidity is beneficial for germination, so use covered seed trays or airtight plastic bags placed over pots. (Various plastic food take-out containers work well for this.) The medium need not be deep for germination; 1 to 2 inches (2.5–5 cm) is adequate.

After stratification, place the containers in bright conditions or close to grow lights. Avoid hot sun or the containers could overheat. The seed will usually germinate in four to eight weeks. Remove the covers at this time. When the plants are of manageable size, they may be picked out using forceps and planted in their permanent pots. Try not to break the small roots.

LEAF CUTTINGS: This is a fascinating way to reproduce sundews, and larger plants can be grown more quickly. This method works for nearly all species. *Drosera regia* and *D. burmanni* are two exceptions, and many tuberous *Drosera* produce too few leaves to be used for cuttings and have such short growing seasons that they can run into problems.

Leaf cuttings work best when taken early in the growing season. Snip off the leaves at their petioles. With the woolly sundews, I've had better success plucking the petiole from the plant with its base intact.

Lay the leaves flat, tentacle-side-up, on a peat and sand mix, long-fibered sphagnum, or milled sphagnum. Using pinches of the soil, secure the leaves at their ends, but do not cover the tentacles to any great extent. Long leaves, such as those of *Drosera filiformis* or the forked sundews, can be cut into segments about 2 inches (5 cm) long.

Keep the cuttings in covered containers to ensure high humidity. Place in very bright light, but not direct sun, or place them under grow lights. After several weeks tiny plantlets will appear, usually from the tentacles. Check the cuttings frequently and resecure them if movement causes the backs of the leaves to lose contact with the medium.

One method I have used to enhance leaf cuttings is to dip them in a solution of SUPERthrive (one drop per cup [235 ml] of water) before laying them out.

Also, many leaves will bud when floated in small covered cups or petri dishes filled with pure distilled water. Change the water weekly or if algae appear.

ROOT CUTTINGS: This method is also easy and can result in mature plants within one season. It works best with sundews that have long, thick roots, such as the cape sundews, most rosetted subtropicals, forked sundews, and *Drosera regia*. Root cuttings work best when taken early in the growing season or at the tail end of dormancy, if the plants have one.

Remove the plant from its soil and cut off healthy roots. These roots are generally black with whitish tips. The mother plant can be repotted. Cut the roots into pieces about 2 inches (5 cm) long. Using covered containers similar

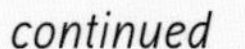
continued

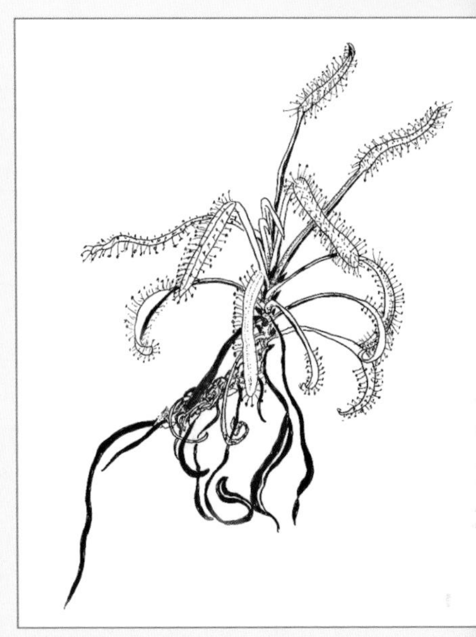

Drosera capensis.

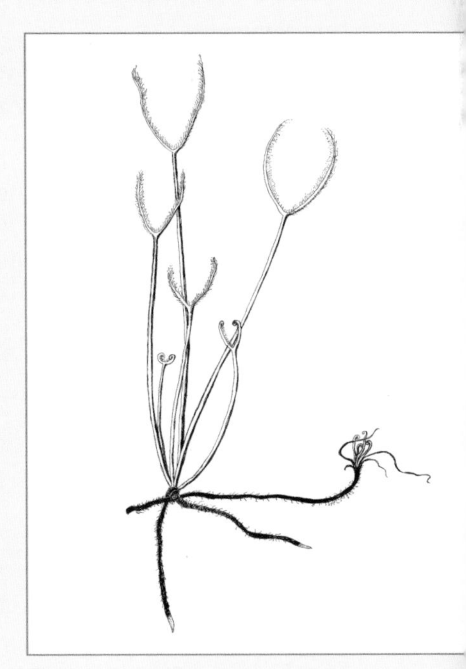

Drosera binata var. *binata.*

to those for leaf cuttings, lay these root pieces horizontally on a layer of medium about 2 inches (5 cm) deep. Cover the roots with no more than ½ inch (1.3 cm) of the medium. Keep the containers covered and in very bright light. After several weeks the plantlets should appear. When a few leaves come up, remove the covers and place the plants in sunnier locations. They can usually be potted up within a few months of sprouting.

GEMMAE: The most reliable way to propagate the pygmy *Drosera* is by the brood bodies they produce in autumn. The only tricky part is removing the gemmae. If the pygmies are in pots, remove the pots from the water tray and allow them to drain for several hours. Then hold the pots upside down over a sheet of plastic or paper. Using a toothpick, tease the crowns of gemmae in the center of each plant. Mature gemmae will pop and shoot off onto the sheet. Collect them with a slightly moistened fingertip. By gently rubbing your finger and thumb together, the gemmae can be sowed like seed onto soil where you will permanently grow the plants. They will reach maturity by springtime.

In situations where the growing containers cannot be overturned, such as bog gardens, most gemmae can be removed gently with a moistened fingertip or small wet paintbrush.

Since pygmies will grow wherever the gemmae are placed, you can be clever in growing them. One lady I know, using a large garden bowl, tediously placed individual gemma with forceps in precise positions. Months later the scores of small dime-size rosettes spelled out the word “sundews” rather neatly!

TISSUE CULTURE: Most sundews are easy to start in vitro with the use of sterilized seed.

The West Australian Pitcher Plant

(Cephalotus follicularis)

Cephalotus follicularis.

One of the most desired carnivorous plants in cultivation, *Cephalotus* is a curious pitcher plant of compact growth with bristly, colorful traps that are seemingly yawning for a bite to eat.

The first published account of this pitcher plant was produced in 1806 by the naturalist La Billardière, who had accompanied the French explorer d'Entrecasteaux on an expedition to southern Australia some years before. A monotypic genus, the Latin name refers to both the structure of the flower as well as the leaf. Another common name for *Cephalotus* is the Albany pitcher plant, named after the town roughly in the center of its distribution.

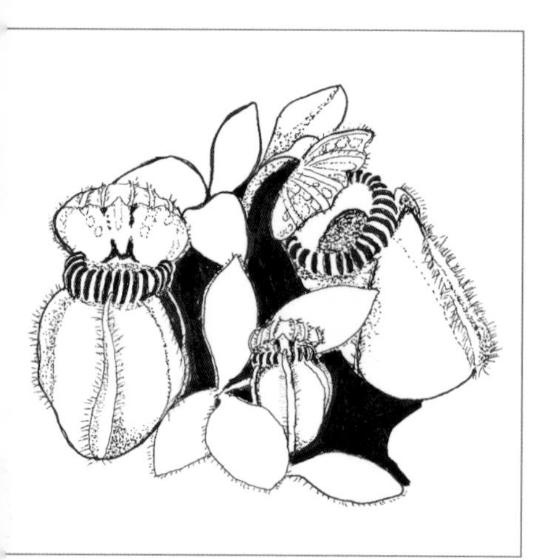

The plants grow naturally in a narrow coastal strip around Esperance Bay in extreme southwestern Australia, a range barely 250 miles (402 km) long. A Mediterranean climate of warm, dry summers and cool, wet winters belies the fact that the plants grow in permanently damp areas on the edges of swamps. As the area is near the coast, the summers are cool, and light frost occurs occasionally in the winter. The plants often grow in clusters under the partial shade of sedge grass in wet, open areas. Sometimes colonies grow only a few yards or meters from saltwater beaches, clinging to mossy banks where fresh water trickles. The plants are often found accompanied by *Drosera hamiltonii*. The soils are peaty sand. Like many CPs, many populations of *Cephalotus* seem to depend on periodic naturally occurring fires in the wild, which clear out brush and small trees, recreating the open, sunny habitats this small species enjoys. However, fire is not necessary for seed germination.

Seedlings grow slowly into low-growing plants that spread through creeping, branching underground rhizomes. *Cephalotus* is peculiar in that it produces two main forms of leaves. As the days grow longer in spring, several noncarnivorous leaves emerge from each growing point, probably to aid in photosynthesis. These are pointed, oval leaves of a rich, shiny green on short petioles. As summer arrives, the pitchers develop, starting as fuzzy knobs at the ends of elongating petioles that slowly inflate and open as a pitcher trap. Occasional third and even fourth kinds of leaves may form, perhaps representing primitive leaves. Some may appear as hollow cups with forked edges, or similar to the foliage leaf but with unusual projectiles emerging from them. Only the pitchers catch insects.

Pool of death! Insects slip from the nectar-baited teeth into digestive juices below. The overhanging white collar prevents their escape.

The plants slow or stop their growth in winter but are evergreen perennials—older leaves and pitchers dying off as new ones form the following year. Flowers usually occur in summer. The stalks are surprisingly long for such a small plant; they may grow to 2 feet (0.6 m) or more, semierect or leaning along the ground. Clusters of the small flowers appear scattered along the upper part. They lack petals, are a pale green, and are less than 1/4 inch (6 mm) across.

The hairy bristles are ladders for insects to climb—a journey they may soon regret.

The pitchers, though small, are elaborately designed. In the typical form of the plant, the traps are 1 to 1 1/2 inches (2.5–3.8 cm) in height.

Reminiscent of an Indian moccasin standing on its toe, the pitchers all face outward in a rosette pattern. Hairy bristles grow on the lid and along three rows of ribs on the pitcher's front. These are no doubt "guide hairs" that lead an insect toward the open mouth. Nectar glands lure the prey—usually small ants—toward a ridge of sharp, incurved teeth. At the base of these teeth the nectar is the heaviest, and insects will often lean over the slippery teeth to get a better taste.

The rain-protecting lid hangs over the mouth of the pitcher much like a canopy. Transparent light windows on this lid are similar to those found on *Darlingtonia* and some *Sarracenia* pitcher plants. But these window panes are not meant to deceptively indicate a way of escape; rather, they allow light to enter the pitcher cavity. The light shines on the otherwise murky pool of digestive juices that lie at the bottom of the leaf, making the depths of the trap more inviting to insects.

Unique among pitcher plants, *Cephalotus* has a bright white collar that overhangs the well of juices. This collar is slippery and is also baited with nectar. Insects hanging on the downward curving teeth may attempt to taste the nectar on this waxy, almost funnel-shaped collar. The prey slips and tumbles into the digestive fluids below.

A large container of hungry *Cephalotus*.

Even if a frantic insect manages to scale the walls of this watery pit, it cannot bypass the overhanging collar and escape. Soon the insect drowns, and it is slowly dissolved in a bath of acids and enzymes.

In the sun, West Australian pitcher plants can develop a beautiful coloration of reds and purples. Shaded, the pitchers may grow larger but remain greenish. Although the lids don't close to capture prey, in drying conditions of low humidity they may slowly fold downward, covering the mouth, perhaps to discourage the evaporation of its digestive juices.

Cultivars

Currently there are two *Cephalotus* cultivars officially registered with the ICPS.

Cephalotus 'Hummer's Giant'
Chosen by John Hummer of the United States, this plant can attain pitchers 2 1/2 to 3 inches (6–7.5 cm) in height on mature specimens.

Cephalotus 'Eden Black'
Registered by S. Morely in England, this handsome plant can develop such a dark hue in high light levels that it appears blackish purple.

There are many other unofficially named cultivars of this species listed on some Internet databases, mostly for size or coloration, such as 'vigorous clumping', 'Giant', and 'Big Boy'. If you are interested in large cultivars, it's interesting to note that Damon Collingsworth has grown typical *Cephalotus* to a very large size at California Carnivores with the occasional application of the foliar fertilizer Maxsea.

CULTIVATION

(See Parts One and Two for further information.)

SOIL RECIPE: West Australian pitcher plants do best in a mix of one part peat to two parts sand.

CONTAINERS: Plastic containers are best, but they also do well in terra-cotta or glazed clay. Always use drained containers; shallow undrained containers will rot their roots. A 4-inch (10 cm) pot suits a single mature plant.

WATERING: Use the tray method. *Cephalotus* dislikes long periods of being waterlogged, so it is best to allow the water in the tray to evaporate before adding more. Curiously, some wild populations thrive on sandy slopes where water constantly trickles; however, in cultivation constantly saturated soils appear to rot the roots.

continued

Cultivation, *continued*

LIGHT: From full sun to part shade. The more sun, the more color in the pitchers.

CLIMATE: *Cephalotus* does best with moderately warm summers and cool to chilly winters. The plants may die in long periods of very hot weather, and they enjoy cool summer nights. They are tolerant of brief light frost down to a couple of degrees below freezing, but they may be killed at lower temperatures.

DORMANCY: *Cephalotus* slow or stop growth during their chilly winter dormancy.

OUTDOORS: A fine potted specimen for partly sunny decks, patios, and porches in Mediterranean and warm temperate climates. Appreciates cool nights in the summertime. Avoid extreme heat for extended periods and protect from freezing.

BOG GARDENS: Does well in deeper bogs that are not waterlogged continuously.

WINDOWSILLS: Does well by a sunny window with fairly high humidity. Keep cooler during the winter rest period.

TERRARIUMS/GROW LIGHTS: Excellent as a potted specimen year-round in the unheated greenhouse-style terrarium. Colorful and vigorous under grow lights. However, reduce the photoperiod and temperature during winter or relocate to a cooler area for its dormancy. Warm conditions year-round will eventually lead to rot and death.

GREENHOUSES: Thrives in cool and warm greenhouses.

FEEDING: Small sow or pill bugs, tiny ants, or dried insects work well. Wingless fruit flies or baby crickets can also be fed to your plant.

FERTILIZING: Occasional misting of foliage can be beneficial.

TRANSPLANTING: It is best to divide and transplant during the end of dormancy.

GROOMING: Trim off old pitchers and leaves as they turn brown.

PESTS AND DISEASES: Aphids are rare on *Cephalotus* due to the bristly nature of the developing pitchers and the tough foliage leaves. The primary pest is scale. Fungi and powdery mildew attack is a sign of low light and stagnant, overly humid air.

PROPAGATION

Although the West Australian pitcher plant is fairly easy to propagate through various methods, it is a slow-growing plant and is still rather scarce and much sought after by hobbyists. Plants offered for sale are most frequently produced through division and leaf cuttings, and they can be expensive. Tissue-cultured plants are problematic, hence their rarity.

SEED: The tiny flowers open on the tall stalk a few at a time. For best seed set, it is wise to use a small paintbrush and tease each flower on a daily basis, transferring pollen from one to another. In several weeks most pods will produce a few hairy, brown fruits—a few seed in each one.

The seed are not viable for long, so if you refrigerate them, do so for no more than two to four months; it's best to sow immediately. Sow on their recommended soil or on milled sphagnum. It is best to stratify the seed, keeping them damp and chilly, for two to three months. With increased light and warmer temperatures, the seed will start germinating after several weeks. Some seed may take a few months to germinate. Slow growing, the rosettes will increase by about 1 inch (2.5 cm) a year.

The flowers of *Cephalotus*.

DIVISION: Use the same method as under the following "Root Cuttings" when your plants have produced several crowns. While dividing these

continued

growing points, you may take some leaf and root cuttings as well. Keep bright and humid as the divisions recuperate.

LEAF CUTTINGS: This is a faster way to grow larger plants more quickly than seed. Remove leaves or pitchers from the rhizome by gently tugging them by the petiole, trying to get as much of the whitish leaf base as possible. Lay these right-side-up on a damp mix of one part peat to one part sand, or use milled or long-fibered sphagnum. Place pinches of medium lightly over the broken base of the petioles to keep the cutting in place. Cover with clear plastic or use a propagating tray. Keep bright, and maintain cool to moderate temperatures. Plantlets will appear usually at the cut end of the petiole in several weeks. After a few months, when a small root system has formed, the plants can be transferred to small pots. A mature plant can be attained this way in about two to three years.

ROOT CUTTINGS: Mature plants can be produced this way somewhat faster than by leaf cuttings. Remove a large plant from its medium in late winter. Rinse off all soil. The thick rhizome and its wiry roots may be used for cuttings (as well as the leaves), or you may simply trim some of these away, leaving much of the mother plant intact, and later repot it. Cut the rhizome and roots into pieces 1 to 2 inches (2.5–5 cm) long. Lay these on the soils recommended for leaf cuttings, but cover them with about $^1/_2$ inch (1.3 cm) of the medium. Keep bright, damp, and covered with clear plastic. Shoots will appear in several weeks. After a few months, separate and pot them.

TISSUE CULTURE: *Cephalotus* can be propagated in vitro primarily through sterilized seed. Removal from flask and introduction into soil can be difficult, and plantlets in flask often produce an abundance of small leaves but unfortunately weak or nonexistent roots. If the leafy growth survives the transition to soil and successfully roots, it can still be slow to attain large size.

The Dewy Pine

(Drosophyllum lusitanicum)

A flowering dewy pine in a large clay pot.

The alluring aroma of honey exuded by the dewy pine is the insidious trick used by this plant to catch flying insects, suffocate them, and eat them.

A most unusual plant, *Drosophyllum lusitanicum* was known by locals as a "flycatcher" for centuries, before studies by Charles Darwin and, later, Dr. A. Quintanilha proved the plant to be carnivorous. Even more surprising, recent DNA studies suggest that this plant (the only species in its genus) may share distant ancestors with *Nepenthes*, the tropical pitcher plants of Asia, and with Triphyophyllum.

What is most striking about the dewy pine is that it is an exception within the carnivorous plant world. Far from being a denizen of a swamp, the dewy pine is native only to limited areas of coastal Portugal, southern Spain, and northern Morocco, where it grows on rather dry hills. The soil is usually sandy gravel caught between boulders and chunky rock, where the most runoff from rain flows. Inhabitants of a Mediterranean climate, dewy pines get most of their rain in winter. Summers can be very hot, although some coastal populations can be cooled by marine-induced fog.

Dewy pines produce several bright yellow spring flowers that scatter their seed in summer on dry soil. When winter rains arrive, the soaked seeds begin to germinate. The seedlings grow quickly, sending their roots rapidly into the gravelly soil. The plants can grow quickly, producing mature-size foliage within a few months.

The flower of the dewy pine.

Dewy pines produce a cluster of narrow, linear leaves around 8 to 10 inches (20–25 cm) in length. Their upright nature resembles a tuft of pine needles, hence their common name. As these leaves die, they produce a skirted thatch around the slowly sprawling, woody stem. Often the plants produce offshoots from their stems, resulting in a branching, semi-prostrate plant that looks like a small, scrubby bush.

The long, thin leaves are highly developed. A shallow fold, or furrow, runs along the length of the leaf on its upper side. The undersurface of the leaf is lined with several rows of stalked glands that are tinted red.

A drop of mucilage, or glue, is secreted by each gland. This glittering drop not only magnifies the red color of the gland but also produces a heavy aroma of bee honey, making the whole plant smell rather pleasantly sweet. Since the leaves are held more or less upright, this sparkling reddishness combined with its promising scent invites insects to alight on its leaves for a taste of nectar.

The unfortunate insect immediately finds itself mired in an oily fluid. Unlike the sticky glue of sundews, this mucilage readily pulls off from the glands, adhering to the struggling insect. Larger, stronger prey such as flies frantically move up or down the leaf, trying to free themselves. As they do, they pull drop after drop of the viscid fluid from the glands. Soon they are overwhelmed, the breathing holes along the sides of their bodies smothered in the clinging liquid. The insect soon suffocates. Tiny prey, like gnats, are usually overwhelmed in a single drop.

Next, sessile glands along the narrow leaf surface secrete digestive juices under the dead body of the prey. Strong enzymes dissolve the soft parts of the insect. This fluid trickles down the leaf, where it is reabsorbed like a soup of nutrients for the plant.

There is a deep, abiding mystery concerning *Drosophyllum lusitanicum* that has yet to be answered. Charles Darwin was the first to notice that the plant continuously secreted its glue, probably to compensate for water evaporation in its dry climate. When he covered the plant in a well-wetted glass bell jar to ensure very high humidity, mucilage was secreted in such quantity that it dripped copiously off the leaves, not being able to evaporate into the already saturated air. More recently, Stewart McPherson and his colleagues, studying the plants in their wild habitats,

The oily glands can overwhelm even the largest of prey.

confirm that even large plants have surprisingly shallow roots penetrating the very dry soil not more than 12 to 15 inches (30.5–38 cm). The baffling enigma is, where does the plant get the water required to produce such large quantities of glue? In its natural habitat, the soils are bone-dry for several months of the year, yet the plants remain dewy. In cultivation, the plants require summer watering or they wilt and die.

CULTIVATION

(See Parts One and Two for further information.)

SOIL RECIPE: The most successful recipe I have used is a mix of one part each of the following: horticultural sand, perlite, pumice, lava rock, and sphagnum peat. Never use more than 20 percent peat. This soil allows for good drainage.

CONTAINERS: A number of years ago I advocated "slack-potting" the dewy pine, a method of cultivating this species using a pot nestled into a larger pot. I no longer endorse this method and instead recommend a single large, unglazed terra-cotta clay pot 12 inches (30.5 cm) or more in diameter. At California Carnivores, I invented a method to allow for safe shipping of younger plants. We start baby dewy pines in small peat pots about 3 to 4 inches (7.5–10 cm) across. The plants grow vigorously in these pots for several months. Our customers insert the peat pots into large clay pots of soil as recommended above. The peat pots decompose and allow the plant's roots to penetrate into the soil of the larger pot. A topdressing of red lava rock or gray pumice can be quite attractive.

A baby dewy pine in a peat pot being transplanted into its permanent home.

WATERING: Dewy pines require a drier soil than most carnivorous plants and should never be kept on the tray system with the pot immersed in water. Always allow good drainage. In greenhouses and other controlled environments, water the plants about once or twice a week most of the year. I like to keep the medium dampish for several days at a time during the winter months. It is crucial that after spring flowering the soil be kept on the dry side for a few months. Only seedlings should be kept wet until they are several months old.

LIGHT: *Drosophyllum* does best in full sun most of the day.

CLIMATE: Dewy pines grow naturally in a Mediterranean climate that is warm temperate. The warm, dry summers usually have cooler nights; the winters are chilly and wet. The plants can survive hot summer days over 100°F (38°C) as well as brief freezes into the 20s°F (-5°C). However, repeated freezes in winter over several weeks can weaken and kill the plants.

Two customers of California Carnivores, living in very different climates and following our cultivation techniques, have grown very healthy *Drosophyllum* outdoors. Jason Herritz lives in the Arizona desert, where summer temperatures hit 110°F (43°C). Burnley Cook lives in Mississippi, with its humid, warm summer nights. Mr. Cook grows his plants on a patio in the sun but protects them from too much summer rain.

DORMANCY: Dewy pines slow their growth during winter but do not go dormant, even after frost.

OUTDOORS: If you live in a Mediterranean climate such as that of California, Western Australia, or South Africa, this is a wonderful potted plant for your deck or patio. Prolonged wet soils in winter are not harmful. You can also grow the plant outdoors in other temperate areas in a sunny place protected from too much summer rain, such as a covered porch. See earlier comments under Climate.

BOG GARDENS: This is not possible for dewy pines, but they should do well in Mediterranean coastal climates as a potted plant submerged in the ground in a drained container. The plants might be able to be naturalized in gravelly soils in places such as coastal California or Western Australia.

continued

Cultivation, *continued*

In these Mediterranean climates, one might dig a hole and fill it with soil (see Soil Recipe, provided earlier) and try a dewy pine in the drier parts of a garden.

WINDOWSILLS: Very doubtful, but possible in sunrooms or solariums that are more like greenhouses.

TERRARIUMS/GROW LIGHTS: I have never known anyone to successfully grow these plants long term in a terrarium, but it might be possible under powerful lighting as a potted plant in a well-ventilated, very large tank, or under lights in the open air.

GREENHOUSES: An excellent candidate for the cold house, cool house, and warm house. Don't overwater the plants in summer. For best health, feed the plants liberally with live or dried insects.

FEEDING: Dewy pines will catch an abundance of flies, moths, gnats, and mosquitoes when grown outdoors. Dried insects applied to the leaves will make your dewy pine drool in appreciation.

FERTILIZING: Not necessary if the plants catch plenty of insects. A foliar feeding of heavily diluted fertilizer can be beneficial, if the plants are not feeding on live or dried insects.

TRANSPLANTING: It is best to introduce baby dewys to their permanent pots when young, without root disturbance. Large plants usually die after root disturbance.

GROOMING: Most growers leave the thatch of old leaves along the stem, but they can be trimmed if you wish.

PESTS AND DISEASES: I have never experienced pests on my *Drosophyllum*, but should any occur I would try a pyrethrin-based insecticide as a method of control, or I would place a flea collar very close to the problem. Diseases are mostly fungus or stem rot, which may affect plants that are kept too wet or are in a humid, low-light, stagnant-air environment.

PROPAGATION

The most frustrating problem with *Drosophyllum* is getting the germinated seedling plants to survive the first few crucial months of their lives. A second difficulty is that currently the only successful way to propagate the plants is by seed. Leaf cuttings, root cuttings, and division all fail. Tissue-cultured plants from seed may be a possibility.

Dewy pines usually flower in spring or early summer, but in cultivation can bloom sometimes as late as early autumn. The handsome, bright yellow flowers are about 1 inch (2.5 cm) in diameter and will self-pollinate on closing. But for heavier seed set, tease the flowers with a toothpick or brush when they open (which is only briefly). Several flowers will open over the course of a few days.

A few weeks later the seedpod cracks open, revealing a number of large, coarse black seed. The pod itself is an unusual cone-shaped structure, opaque tan.

I prefer to store the seed dry in the refrigerator until autumn. Scarification of the seed is helpful, so I usually rub the seed firmly between sheets of sandpaper until the hard surface is lightly scraped. Next, I soak the seed in water for a few days.

Sow the seed, without burying it, in a small tray of shallow sand. Keep it damp to wet, and seed should start germinating within days to weeks.

Check daily, and immediately on germination (or at least within a day or two) gently remove the sprouting seed and lay it on the surface of the soil in its permanent pot. Keep the soil damp for a few weeks as the seedling establishes itself. Then begin to allow the soil to dry out between waterings. The plants will grow rapidly; keep them in a sunny location. I have found that when seedlings are growing in more stagnant air, such as in a greenhouse, they are very prone to botrytis or damping-off disease that even fungicides can't seem to control. Our newly potted seedlings thrive much more happily outdoors, even tolerating rain and temperatures down to the frost level.

The plants will be of flowering size in one to two years. However, if the plants are precocious and attempt to flower during the first year, I prefer to cut the emerging buds off to allow development of the plant itself. A heavy diet of insects strengthens the plants, and outdoors they are voracious feeders.

The Rainbow Plants

(Byblis)

Byblis liniflora at sunrise.

In mythology, Byblis was the granddaughter of Apollo; she fell in love with her twin brother, and when he rejected and fled from her, Byblis wept bitter tears and turned into a fountain.

Glittering and often delicate, *Byblis*, the rainbow plants, can sometimes appear as frosted sprays of water, and in sunlight they can sparkle with multicolored hues. Their shining leaves and pretty flowers mask their deadly nature, as they catch and kill countless tiny insects that make the fatal mistake of alighting on them.

Specimens of *Byblis* plants were discovered a few times from the last decades of the 1700s into the following century, primarily by naturalists on Cook's voyage and the famous Australian botanist James Drummond. For well over a century only two species were accepted: a larger perennial plant, *B. gigantea,* from southwestern Australia, and a smaller, annual species, *B. liniflora,* from the more tropical areas of northern Australia and southern New Guinea. By the end of the twentieth century, Allen Lowrie and his colleagues had separated the two species into seven, based mostly on differences in flower structure. (See the later discussion of the annual *Byblis* species, page 227.)

In 1905, A. M. Bruce fed plants of *Byblis gigantea* tiny bits of hard-boiled egg and, when the egg dissolved in a few days, concluded that the plants were carnivorous.

Because of their flower structure and glands, rainbow plants were at first assumed to be related to butterworts and bladderworts of the family Lentibulariaceae, but finally were given their own family, Byblidaceae. Recent DNA research by Steve Williams has shown that their closest relatives are indeed the butterworts, found far from the Australian continent.

The seven species all have similar structures and trapping mechanisms. The plants have narrow stems that sometimes branch and trail along the ground or lean against and climb nearby grasses or other plants. The leaves are fine and linear, radiating in all directions from the stem. The whole plant is covered with two types of glands. The first are stalked, hair-like glands capped with a clear, sticky glue. Small insects like gnats and mosquitoes become trapped in the glue and struggle to escape. Eventually these insects die from exhaustion or are suffocated.

It is then that the second glands come into action. These sessile glands lie flat along the leaves, and they secrete the digestive juices that dissolve the soft parts of the prey.

During the twentieth century occasional research into these secreted juices detected no measurable enzyme production. But by 2005, research by Siegfried Hartmeyer and others finally found enzymes in at least two species. Furthermore, much like the controversial *Roridula* plants of South Africa, rainbow plants play host to assassin bugs, and the plants may benefit from the bugs' secretion in the same way. These predatory bugs can mysteriously live on the sticky leaves of *Byblis* without being trapped. They suck the juices of helplessly caught insects, and most likely the plants can absorb the fecal secretions of the bugs, much as *Roridula* does.

The Perennial Species

Byblis gigantea and *B. lamellata* grow in regions of southwestern Australia in areas surrounding Perth. The climate is Mediterranean, with cool, wet winters and hot, dry summers.

Byblis gigantea in the wild during Western Australia's wet season.

The plants are perennial, and the seedlings grow rapidly, sending roots deep into the sandy soil. The seeds usually germinate when the rains of late autumn begin, but they need a quick-moving brushfire in summer to chemically condition them to germinate. In the wild, seeds may lie on the ground for several years before a fire moves through. But during the rainy season subsequent to a fire, hundreds of seeds will sprout. The seedlings can grow so fast that they can be flowering by the following spring, just before the onset of the hot, dry summer.

These two species of giant rainbow plants are really no larger than a small bush. Stems grow from a woody base 1 or 2 feet (0.3–0.6 m) tall. The linear leaves average about 10 inches (25 cm) long. The handsome flowers are more or less flat faced, deep pink to violet blue, and about 1 inch (2.5 cm) across, on stalks that look similar to the leaves. To release their pollen usually requires the vibration of insect wings—the pollen sprays onto the insect in a little cloud. The insect then deposits the pollen onto the next flower.

Byblis gigantea and *B. lamellata* usually grow on the sandy edges of wet winter swamps. When summer arrives and the soil dries out, the stems of the plants die down to the woody base, and the plants remain dormant until the rains return the following autumn. However, both species grow year-round in cultivation and in some areas where the soil is always damp.

Both perennial species can live for many years, if not for decades. Aside from minor differences in flower structure, both plants are very similar. *B. gigantea* has short stems with leaves clustered at the growing point; *B. lamellata*, found in the northern regions of the genus's distribution, has taller stems with leaves more sparsely arranged along them.

The Annual Species

For nearly two centuries all of the annual rainbow plants were considered to be one species, *Byblis liniflora*. In recent decades, however, Allen Lowrie and his associates have separated the species into several, based mostly on technical differences of their flowers: *B. liniflora, B. aquatica, B. guehoi, B. filifolia*, and *B. rorida*.

They are found in northern Australia, from Western Australia through the Northern Territory and into the Cape York Peninsula in Queensland. *Byblis filifolia* is also found in southern New Guinea. The climate is tropical and warm all year but is marked by a monsoonal wet summer with a drier winter.

Plants of *Byblis* typically grow in sandy areas bordering summer-flowing streams and water holes. The small, poppy-seed-like seeds germinate rapidly with the warm rains, and the plants grow very fast, flowering sometimes within weeks of germination.

The plants are small, delicate, and scrambling, their stems approaching 12 inches (30.5 cm) in length and sometimes longer. *Byblis rorida* and *B. filifolia* can have stems approaching 4 feet (1.2 m) in length. The leaves are threadlike and 5 to 6 inches (12.5–15 cm) long. The whole plant is densely covered in sticky hairs, and in sunlight can put on a rather spectacular show, glittering like frosted miniature Christmas trees.

The flowers are produced profusely during the life span of the plant, with many small, 1/4-inch (6 mm) blooms hovering amid the leaves.

Byblis aquatica growing in shallow water after seasonal rains. The seedpods open when waterlogged.

In the most commonly grown species, *B. liniflora*, the flowers are the exact color of amethyst and will open in sunlight over several days. They self-pollinate before withering. *B. aquatica* usually also self-pollinates, but the other species appear to require insect vibration to release pollen and then be crossed with a separate flower of a genetically different plant. The flower colors of the species other than *liniflora* are almost invariably purple, violet to mauve; however, occasional white-flowered forms are also known. In *B. filifolia* the flowers can surpass 1 inch (2.5 cm) across.

Cultivated *Byblis filifolia* in mass flower at Isao Takai's greenhouse in Japan.

Isao Takai, a carnivorous plant enthusiast in Japan, is well known for his amazing success in growing rainbow plants in his hothouse. Giant clones of *B. filifolia* that he has grown can be so smothered in pink and white flowers that the threadlike leaves are barely visible amid the hundreds of blooms. However, he has also reported an unusual occurrence among his plants. Normally, large insects like grasshoppers are nearly impossible for delicate-leaved *Byblis* to catch, but Isao has reported that such insects, moving about in the mass of sticky leaves, appear to become drugged and paralyzed and eventually die. Additionally, Brian Barnes, a grower in Florida, reports that if he shakes specimens of *B. filifolia*, they immediately emit a fungal aroma. The possibility that the plant somehow immobilizes insects through some sort of chemical fumes still needs to be explored.

Cultivars

Brian Barnes has registered two cultivars of *Byblis*.

Byblis 'David'

A form of this genus, of dubious ancestry, in which the leaves and flower stalks slowly move from an upright position downward, offering support like tripods for the trailing stem. There are also sessile glands where the leaves join the stem. An annual, the flowers self-pollinate and can be used to perpetuate it as a cultivar.

Byblis 'Goliath'

Grown from *B. filifolia* seed. It produces tall branching stems to 3 feet (0.9 m), and the flower stalks bend downard after pollination, similar to the leaves of 'David', a phenomenon known as a "pulvinus" (plural "pulvini"). Brian suggests that the fungal aroma of a shaken plant may repel herbivores. This cultivar is easily reproduced through stem cuttings.

CULTIVATION

(See Parts One and Two for further details.)

SOIL RECIPE: All species do well in a mix of two parts sand to one part peat. You can also add an extra part of perlite, lava rock, or pumice.

CONTAINERS: Plastic works well, but if you use fire to germinate seed of *B. gigantea*, use clay or glazed ceramic. Use 4- to 6-inch (10–15 cm) pots for

continued

Cultivation, *continued*

the annual species (the plants can live longer in deeper pots), 6- to 10-inch (15–25 cm) pots for the perennial.

WATERING: Use the tray method, but allow the water to evaporate before replenishing. Keep the soil just damp and not permanently waterlogged, except for *B. aquatica*, which enjoys wetter conditions.

LIGHT: Full to partial sun for both species.

CLIMATE: The annual species do best in tropical climates; their shorter life span allows you to grow them during any span of several months of warm weather. The perennial species do best in Mediterranean climates that are almost frost free, with cool summer nights, but they may succeed in subtropical climates.

DORMANCY: Species like *B. gigantea* die down to their stems during dry dormancy in the wild, but in cultivation they usually do not go dormant when kept wet. Annuals like *B. liniflora* die after one or two years.

OUTDOORS: The perennial species do well in almost frost-free Mediterranean climates and probably warm temperate to subtropical climates as well. The annual species will usually grow well with several months of warm weather, or in tropical areas.

BOG GARDENS: Excellent in bog gardens. Best in higher areas of the bog. Pay attention to climate requirements of the species (described earlier).

WINDOWSILLS: I have seen only healthy *B. liniflora* do well on very sunny, warm windowsills, but other species may work. Direct sun would be a requirement.

TERRARIUM/GROW LIGHTS: The smaller tropical annuals do well in warm tanks under lights, but the larger *B. filifolia* will grow too large for a terrarium; it can be grown out of a tank under high-intensity grow lights. This is true also for the larger perennial species, which also prefer cooler night temperatures than the tropical forms. Excellent for *B. liniflora*.

GREENHOUSES: The perennial species thrive in cool and warm greenhouses. The tropical annual species do best in warm houses, hothouses, and stove houses, but can be grown from late spring until autumn in other houses, too.

FEEDING: Small insects such as gnats and fruit flies are appreciated.

FERTILIZING: A light monthly misting with fertilizer helps.

TRANSPLANTING: Any root disturbance can shock and kill the annuals. The perennials can be successfully transplanted when necessary. Soak the rootstock in a vitamin B1 solution to overcome shock.

GROOMING: I like the look of the old skirt of leaves, which also support the growing point, but trimming can be done.

PESTS AND DISEASES: On very rare occasions, mealybugs or aphids may attack the perennial species. Damping-off fungus can be a problem with seedlings; treat with a fungicide. Snails and slugs like the baby plants.

PROPAGATION

SEED: *B. liniflora* is simple to propagate by seed. Although the flowers will usually self-pollinate, I like to tease them with a toothpick to ensure good seed set. A few weeks later, the brown pods will split, revealing up to two dozen seed. Store these dry in the refrigerator and sow in spring. Or, after a rest period of a few weeks, the seed will germinate in any warm, bright environment, such as under grow lights. Simply scatter the poppy-seed-like seed on the surface of the preferred soil mix and keep it warm and damp.

Typically, *B. liniflora* will die off in the winter months, but in terrariums and sometimes on windowsills they can live two years or longer. Apparently only *B. liniflora* and *B. aquatica* can be self-pollinated. All other species of rainbow plants, whether annual or perennial, require pollen to be transferred to another flower of a different genetic clone of the same species. (This means you will need two different seed-grown plants.) Additionally, *B. aquatica* grows in areas that become flooded with standing water, and perhaps because of this, the seed capsules won't open until wetted.

continued

The flower of *Byblis gigantea*.

Propagation, *continued*

For all species except *B. liniflora*, fire or smoke treatment is usually necessary for seed germination. The following pollination and seed treatment for *B.gigantea* usually works for all other species.

The flowers of *B. gigantea* can be tricky to pollinate. I have found the tuning fork method to be the best. Touch the vibrating fork to the center of the flower while holding a sheet of paper under the petals. If the pollen is ripe, a cloud of pollen grains will spray onto the paper. Collect this with a small paintbrush. The pollen must be dabbed onto the stigmas of a separate flower, of a genetically different plant, as self-pollination seems to fail. Seed will be produced in several weeks.

It is believed that the chemicals released by smoke are what triggers seed germination in *B. gigantea*—not the heat of fire. There are several methods used to achieve this in cultivation.

Lightly press the seed onto the damp soil surface in a clay or ceramic pot. Place a small loose pile of dried grass, hay, or paper towels over the soil. Light with a match and gently blow on the fire to ensure a good, smoky blaze. When the fire burns out, allow to cool and gently blow off excess ash. Lightly sprinkle with water. Germination should follow in a few weeks.

Alternatively, sow the seed in the same way and place a burning cigarette on the soil. Cover the pot with a clear plastic bag and allow the smoke to become dense. The cigarette will usually extinguish itself before too long. Leave the bag on the pot until the smoke eventually clears.

Liquid smoke, available at grocery stores and used to flavor meats, is a much easier treatment. I suggest soaking the seed in this product, or try placing a few drops of the liquid onto your sowed seed.

Smoke disks have recently been developed to enhance seed germination of certain plants. These paper disks are soaked in water along with the seed. To find a distributor, do a web search for *Kirstenbosch Instant Smoke Plus Seed Primer* or *Kirstenbosch smoke disks.* Finally, gibberellic acid can be used to germinate *B. gigantea* seed. It is available through companies specializing in science and biological supplies. Gibberellic acid is a powder you dissolve in water at a ratio of about 1 part acid to 1,000 parts water. To achieve this ratio, I find that about 1/8 inch (3 mm) of the acid on the end of a toothpick dissolved into about 1 cup (235 ml) of water works well. Soak the seed for twenty-four hours in this solution before sowing.

DIVISION: Larger plants of *B. gigantea* and *B. lamellata* can be divided. Remove the plant from its pot and rinse away most of the soil. The woody rhizome will usually have multiple growing points. Cut these apart with a sharp knife, trying to ensure that each piece has a few roots attached. Soak in a vitamin B1 solution, then repot the divisions.

ROOT CUTTINGS: Root cuttings can also work. Cut the oldest, thickest roots into pieces about 2 inches (5 cm) long. Lay these on their preferred soil mix and lightly cover with pinches of soil. Keep damp and humid, as in a propagating case, in bright light. Plantlets usually appear in several weeks.

B. filifolia will often produce multiple branching stems. You can cut these branching growing points from the main stem and insert them into long-fibered sphagnum moss, and they will root. After rooting, transfer the new plant to its preferred peat and sand mix.

The Butterworts

(Pinguicula)

"Well, it's a cross between a butterwort and a Venus flytrap."

—Seymour Krelboyne to Mr. Mushnick, when asked where his person-eating plant, Audrey Junior, came from (*Little Shop of Horrors,* 1960)

Pinguicula vulgaris.

A customer visiting my nursery once described the butterworts as "the Shirley Temples of the carnivorous plant world." I must disagree. Butterworts are Patty McCormacks (in honor of the sweet, pretty child actress who played Rhoda Penmark, a devilish girl who enjoyed pushing little old ladies down stairs, in *The Bad Seed*). The fact that Seymour Krelboyne, in the original production of *Little Shop of Horrors*, admits that one of the parents of Audrey Junior was indeed a butterwort boggles my mind. I can only imagine Charles Griffith, the screenwriter, perusing his encyclopedia under "carnivorous plants" and suddenly exclaiming, "Butter-wort! Now that sounds pretty disgusting!" And thus history in sci-fi horticulture was made.

As for real history, butterworts were brought to the attention of Charles Darwin in the early 1870s. A gentleman named Mr. Marshall mentioned to the carnivorous-plant-obsessed Mr. Darwin that he noticed many small and struggling insects on the leaves of butterworts growing in England. Darwin investigated, and sure enough discovered another genus of insect-eating plants.

S. Jost Caspar, in the mid-1960s, wrote a large monograph on butterworts that accumulated all the information on the genus known up to that time. A botanist in East Germany, Herr Caspar was amazed to learn, when the Berlin Wall fell in the late 1980s, that some of the plants had achieved windowsill popularity in the outside world. More recently, much information has been gathered by the studies of Dr. Donald Schnell on the U.S. species, and particularly by Jürg Steiger in Switzerland.

Pinguicula means "little greasy one" in Latin. While butterwort is the common name, more recently some growers have begun to call them "pings." There are around one hundred species known, many of them rather recent discoveries. Pings grow throughout much of the northern hemisphere, from the Arctic Circle down through Siberia, Europe, and North America. They reach their climax in Mexico, where the most spectacular forms exist. A few more species exist down through South America.

Butterworts are typically small, herbaceous plants 2 to 4 inches (5–10 cm) in diameter. Like many other CPs, they grow in a rosette fashion. The leaves of almost all species are flat, with slightly upturned margins. The leaves arise out of the center of the plant, and as they mature they press themselves rather firmly against the ground, although in a few species the leaves may be held semierect or arching. The leaf shape of butterworts may be narrow and tapered to a point or rather oblong to nearly oval. Only a few unusual forms have leaves nearly filiform, and in some species the upturned margins may be lacking or even curved downward. The color of the leaves is almost invariably pale to dark green, but some sun-growing pings may have leaves that turn bronzy or reddish. The roots are usually few and short.

As their Latin and common names suggest, the leaves of butterworts have a distinctly buttery or greasy feel to them. At a glance, the leaves appear perfectly innocent. But when examined closely with a magnifying lens, the surface of the leaves is revealed to be covered with thousands of minute, nearly transparent glandular hairs. The gland at the top of each hair produces a small drop of sticky glue. A second kind of gland also populates the leaf surface. These are flat, sessile glands, and in the right angle of light, when viewed under a lens, they appear as shallow depressions, like small, dished craters. These sessile glands are dry—until prey is caught.

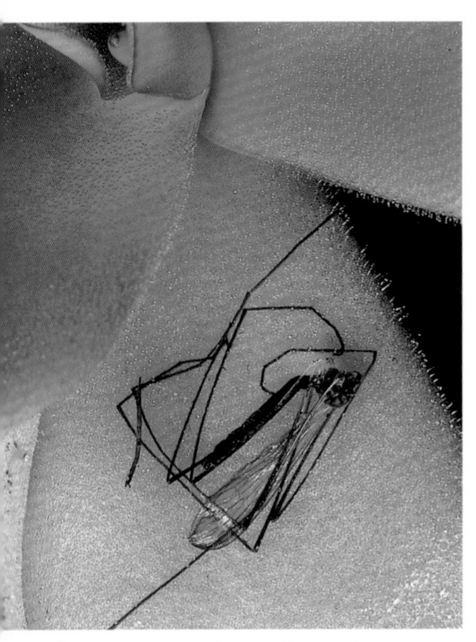

Butterworts only rarely catch prey as large as this unfortunate crane fly.

The stalked glands can catch sunlight, and thus the leaf surface can glitter or shine, sometimes producing a shimmering rainbow effect that may be a lure for prey.

Their victims are generally very small insects such as gnats, springtails, and fruit flies. Rarely can pings catch anything approaching the size of a housefly, although sometimes this does occur. The strongest prey I have ever witnessed caught by a cultivated butterwort was, to my amazement, a newly hatched praying mantis, and in the wild, rather powerful carpenter ants.

When an insect alights on a butterwort leaf, it immediately realizes its mistake: it is mired in the glue of the stalked glands. As it struggles to break free, it pulls more and more drops of glue from the glands until it is overwhelmed and hopelessly stuck. The strength of this glue is powerful—the panicky insects often leave behind twitching legs in their attempt to escape. Minute insects like springtails can be overcome by one or two stalked glands.

It is then that the sessile glands come into play. Almost immediately after the insect is caught, these glands begin to secrete a liquid of acids and enzymes. Soon the dying or dead insect is wetted down by this secretion and digestion begins. The soft parts of the prey dissolve, and the victim is reduced to a pulpy mass, sometimes in a matter of hours. The secretion of digestive fluids can be so copious it can sometimes be seen trickling down the leaf. The sessile glands soon reabsorb this fluid, now rich in nutrients.

That butterworts produce a strong bactericide was noticed quickly by Darwin. Parts of dead insects and bits of meat he fed to his pings that were not wetted by the digestive juices were soon attacked by bacteria. The digested food was bacteria free.

Three forms of *Pinguicula moranensis.*

In fact, this antibacterial property has long been known by northern Europeans. For generations, butterwort leaves were applied to the sores of cattle to promote healing. The leaves were also used to produce a unique curdling effect in goat's milk to produce a ropy, yogurtlike cheese.

Some species of pings also have the power of movement. This is most often seen in species from temperate climates; it is almost entirely lacking in tropical forms. Over a period of a day or so, after the capture of substantial-size prey, the margins of the leaf, already upturned, may curve inward or over the precious food. This has nothing to do with capture but is believed to be helpful in preventing the digestive fluids from drooling off the leaf. Another possible explanation is that it helps prevent the victims from being washed away by the rain. Many butterworts can even "dish" their leaves under prey, giving their juices a convenient place to pool.

The most wonderful thing about *Pinguicula* in cultivation is their flowers. Most of the species in this genus have flowers that are pretty, but the flowers of the tropical forms are extraordinary; they can rival orchids or African violets in their beauty and brilliant color. Many put on quite a beautiful show in the springtime, and some may be in bloom so often and for so long that they can provide many months of pleasure.

The flowers of butterworts may be funnel shaped, cupped, or flat faced, with long or short spurs. The petals appear as two lips: the upper lip divides into two lobes and the lower lip divides into three. In the throat of a butterwort flower are many hairs, often called beards, which in some species may be rather pronounced. During their flowering season, pings usually send up a succession of individual blooms on single stalks.

Since pings grow on nearly half of our planet, their climates obviously differ greatly. In the north, their habitats are often frozen or frosty in winter, and they can sometimes be found accompanied by sundews and pitcher plants. But in the more tropical countries of Mexico and the Caribbean, butterworts have adapted to survive long periods of drought. Some, found growing in alkaline dry areas of Mexico, are even companions of cacti and other succulents. There are also species that grow epiphytically on trees (as do the *Tillandsia* air plants), and one species has been found growing in mossy patches on glaciers in the Arctic.

For convenience, this chapter is divided into three sections, based on the pings' habitats. Specifics on their cultivation and propagation can be found on pages 261 to 266.

Temperate Pings

Butterworts from temperate climates survive long, cold winters by dying down to small, cone-shaped hibernacula. These "resting buds" of tightly held leaf scales often lose their roots in winter and are easily moved about by water at this time, which aids in their distribution. The hibernacula also produce, around their bases, gemmae, or brood bodies, which look like miniature resting buds. Scattered about, they grow into new plants and are very helpful in propagation.

The dormant winter bud of *Pinguicula longifolia* with gemmae around its base and brown dead leaves from the previous season.

Temperate butterworts come into growth in spring, producing flat, often starfishlike rosettes of narrow, tapered leaves with strongly upturned margins. Spring is also their flowering time, and several blooms arise singly on stalks 2 to 3 inches (5–7.5 cm) tall. The flowers are typically cone shaped with long, funnellike spurs. The petals are usually short and most often colored purple, violet, or white, with hairy beards at the entrance of the flower throat.

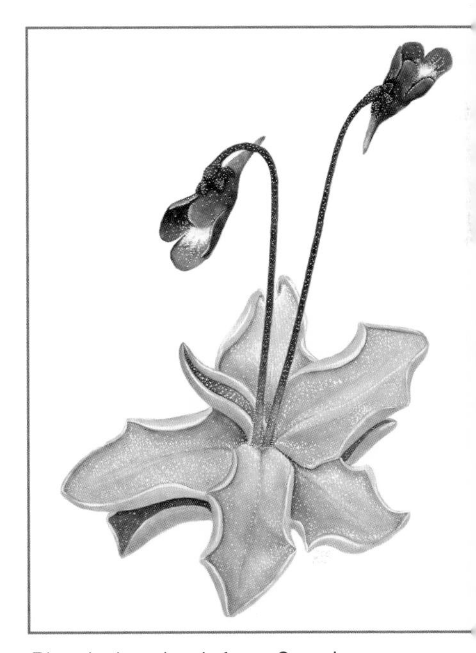

Pinguicula vulgaris from Ontario, Canada, painted by Scott Bennett.

Temperate pings can be found in boggy moors of acidic, peaty soils in North America and Europe, but most often they grow on wet, dripping, gravelly cliffs and grottos, often amid mosses and ferns. In areas such as the Great Lakes region they can be found in flat, damp, rocky soils at the water's edge, sometimes accompanied by *Sarracenia purpurea* ssp. *purpurea* and *Drosera rotundifolia*. Often the soils that temperate pings grow in are more neutral to alkaline than acid. The habitats in which they grow are usually sunny, but the butterworts are often partly shaded by grasses, ferns, and other low-growing vegetation. The water trickling through their gravelly soils is usually cool groundwater.

Pinguicula vulgaris

This is the species Charles Darwin studied, and it is very widespread, growing in North America, Europe, and northern Asia. It likes rocky areas and is often found around waterfalls and lake margins, growing in both acid and alkaline soils. The rosettes are up to 4 or 5 inches (10–12.5 cm) across, of greenish yellow leaves that appear narrow due to their strongly incurved margins. The flowers are about 1 inch (2.5 cm) long, violet with a white throat. There are a few geographical forms from Europe. *P. vulgaris* f. *bicolor* has petals that are white and purple; *P. vulgaris* f. *albida* has white petals; *P. vulgaris* f. *alpicola* has flowers twice the size of the typical species.

Pinguicula macroceras ssp. *nortensis.*

Pinguicula macroceras

This species was once considered a variant of *Pinguicula vulgaris* but is now considered a separate species. It grows in western North America, Japan, and Russia. The flowers are more opened than those of *P. vulgaris*, and the lower lobes are longer. *P. macroceras* ssp. *nortensis* is found in northwestern California and southwestern Oregon. Its flower has an elongated lower-center lobe, and plants growing in full sun in serpentine gravel, often with *Darlingtonia*, can have leaves with a rich chocolaty red coloration.

Pinguicula grandiflora—grand flowers indeed! One of the easiest temperate butterworts to grow.

Pinguicula grandiflora

This is a lovely species when in flower and much easier to grow than the preceding two species. Native to Europe, it is found in hilly, mountainous regions in Ireland, France, Switzerland, and Spain. The foliage is rather similar to that of *P. vulgaris* but can be larger. The beautiful flowers are up to 1½ inches (3.8 cm) long, with broad lobes of rich violet and deep purple veins. *P. grandiflora* f. *pallida* has pale bluish petals with a purple-ringed white throat. *P. grandiflora* ssp. *rosea* has pale, rosy-colored flowers.

The flower of *Pinguicula longifolia* ssp. *caussensis*.

Pinguicula longifolia

This is another easy temperate species that does not require the frigid winters of *P. vulgaris*. It grows in the Pyrenees of southern France and northern Spain. The foliage is similar to that of *P. grandiflora* and *P. vulgaris*. The flowers of *P. longifolia* ssp. *P. longifolia* have long spurs, with pale violet petals, a white throat, and hairy lower lobes. *P. longifolia* ssp. *caussensis* has very pale flowers and a broad lower-center lobe. *P. longifolia* ssp. *reichenbachiana* has small flowers with a prominent, hairy white throat.

There are many other temperate butterworts that require rather similar cultivation techniques, such as the following:

Pinguicula corsica

I enjoyed growing this Mediterranean island species for several years and found it similar to *P. grandiflora* and *P. longifolia*. The flowers have rounded lobes of very pale violet with darker veins.

Pinguicula alpina

A small plant, barely 2 inches (5 cm) across, this species is common in the mountains of Europe and can also be found in lower elevations in Scandinavia and Scotland. The leaves are short and triangular, and the handsome flowers are white with bright yellow throats. When dormant, this species does not lose its roots—so transplant it with care.

Pinguicula ramosa
From the mountains of northern Japan, this rare species has small spatulate leaves. Most unusual are the white flowers, which can appear in twos and threes on a stalk. Almost extinct, this plant has appeared on a Japanese postage stamp. Only two colonies remain.

Pinguicula villosa
This tiny species grows in the Arctic regions of Asia, northern Europe, and North America. I have never known anyone who has grown it, but success might be had by keeping the plants for nine months or so in your freezer (unnecessary, of course, if you live someplace like northern Alaska!). These tiny plants are under 1 inch (2.5 cm) in diameter and grow only a few reddish leaves in their short season. The minute flowers are white to pale purple, with yellow dots on the lower lobes.

Pinguicula vallisneriifolia.

Pinguicula vallisneriifolia
An unusual species from southern Spain. The summer leaves are very narrow and almost 8 inches (20 cm) in length. The flowers are pale bluish purple and the lobes are rather rounded. This species usually grows on shady wet cliffs and has a rather spidery appearance. Unlike other species, *P. vallisneriifolia* grows stolons to produce new plants.

Warm Temperate Pings

There are a number of butterworts that do not form winter resting buds because their native climate is not severe enough to require it. I will refer to these as warm temperate even though some grow in temperate or subtropical conditions. Most of these species prefer permanently wet, acidic soils and can take various degrees of light frosts. Hard freezes can kill them, but even when this occurs in the wild, plants survive from seed.

Many are short-lived perennials, dying off after a couple of years, and at least one is an annual. Like most other butterworts, they usually grow in sunny locations lightly shaded by surrounding vegetation.

The largest group of warm temperate butterworts grows in the southeastern coastal plain of the United States, and most share their habitats with *Sarracenia, Drosera, Dionaea*, and *Utricularia*. This unfortunately does not mean they are as easy to grow as their companions. These species usually grow best in outdoor bog gardens in suitable climates, or in cool and warm greenhouses. It is a good idea to keep up propagation of these plants if you wish to keep them in your collection, since most are short-lived. Most of these have pretty flowers that can be self-pollinated for seed.

Pinguicula caerulea

The violet butterwort is found from the coastal plain of North Carolina south into southern Georgia and most of the Florida peninsula. It forms a handsome rosette of pale green leaves up to 4 inches (10 cm) across. The leaves are oval with strongly incurved margins, making them appear narrow and pointed. The 1-inch-long (2.5 cm), funnel-shaped flowers are usually violet with many purplish veins and a yellow, protruding beard. Some variants are pure purple. The lobes are deeply incised, giving the appearance of many petals. This species is probably the easiest to grow of the warm temperate pings and can survive for many years. It is excellent for a bog garden.

Pinguicula lutea in the wilds of the Gulf Coast.

Pinguicula lutea

This species looks identical to *P. caerulea*, except that the flowers are a bright sulfur yellow. Its range is also similar, except it extends along the Gulf Coast almost to New Orleans. I have found it difficult to maintain long term in cultivation. Usually the plants flower for many months while the plants wither away.

The sulfur-yellow flower of *Pinguicula lutea*.

The primulalike flowers of *Pinguicula primuliflora*.

The flower of *Pinguicula caerulea*, more purple than veined in this form.

Pinguicula primuliflora

This is a popular species in cultivation; in nature it is found in coastal areas of the Florida Panhandle west into Louisiana. The leaf rosettes are similar to those of *P. lutea*, but appear narrower. The plants are often found in mostly shaded areas of very wet peat and sphagnum, often along the edges of streams and ponds. While the plants are short-lived and prone to rot, they spread by producing plantlets at the tips of their leaves, which is helpful for propagation. Keep them shaded from hot sun. The beautiful flowers have incised petals that are pinkish violet with white centers and a yellow beard.

Pinguicula planifolia

This large species is unique. Its 6-inch-wide (15 cm) rosettes of pointed, lance-shaped leaves turn a purplish red in sunny locations. Like *P. primuliflora*, it enjoys wet, peaty soils and is sometimes flooded by shallow water. The flowers are attractive, pale violet, and seem many-petaled due to the deeply incised petals with a protruding yellow beard. Unfortunately, it is difficult to maintain long term, although it fares best in sunny, wet bog gardens. Occasionally it produces leaf buds late in the season. It is found from the western Florida Panhandle to Louisiana.

Pinguicula pumila

A diminutive butterwort rarely approaching $^{3}/_{4}$ inch (2 cm) across, this tiny ping grows in the coastal plain from the Carolinas, south through the Florida peninsula, and west into Texas. It is a pretty miniature that is fairly easy to grow and often self-seeds. The tiny flowers can be purple, yellow, or pink. I grow a pretty blue form from Georgia.

The sticky leaves of *Pinguicula primuliflora*.

Pinguicula ionantha

This severely endangered species was first described in 1961, and early in the twenty-first century the first tissue-cultured plants became available in horticulture. I have found it surprisingly easy to grow. The rosettes approach 4 inches (10 cm) across, the margins of the leaf barely rolled along the sometimes kinked margins. It is fond of very wet areas. The petals are usually white, nearly oval, and indented at their ends. A prominent yellow beard protrudes from the throat. Amazingly, one seed-grown plant we've grown at California Carnivores has purple flowers.

In other parts of the world are a few butterworts that also maintain their leaves year-round.

Pinguicula planifolia.

Pinguicula lusitanica

This is another tiny species that often behaves like an annual but produces abundant seed and is very easy to grow. Its habitat is wet, peaty areas of coastal Europe from England to Spain and northwestern Africa. The rosettes are usually around 1 inch (2.5 cm) across, sometimes larger. The elliptical leaves are deeply rolled, translucent, often pinkish, and veined in red. Hairs along the center of the leaf force insects (often springtails and gnats) to be trapped along the glandular margins. The small, funnel-shaped flowers are pale pink. *P. lusitanica* grows well

The shadows of insect prey are visible through the leaf margins of *Pinguicula lusitanica*. Botanists now believe this diminutive European butterwort is an archaic species.

in cool and warm greenhouses, in terrariums, on windowsills, and in bog gardens. Since it grows quickly from seed, it succeeds as a summer annual outdoors in cold temperate climates, but you should collect seed to sow the following year.

Pinguicula crystallina
Once considered two species, this plant has been combined into two subspecies. My one attempt to grow *P. crystallina* ssp. *crystallina* from Cyprus failed. The flowers are rose to pale blue, with leaf rosettes to 2 inches (5 cm). The subspecies *hirtiflora*, from southern Italy, has stunning large flowers: purple, white in the center, with a bright orange throat. Pollinate them for seed production.

Pinguicula antarctica
This species is the most southern-growing ping, coming from coastal, marshy areas at the southern tip of South America. The cool, wet climate where this species is found was once described to me as Seattle-like. The oval leaves are oblong and pointed, in rosettes rarely approaching 2 inches (5 cm) across, with small lavender flowers. I was able to grow this rare species successfully in my cool house. Pollinate for seed production.

The following butterworts are still rare in cultivation, and I have not had the opportunity to try them, but I suspect they would be successful in the cool house. All are from South America but grow at rather high elevations, so cultivation can be challenging.

Pinguicula elongata
An intriguing species from the highlands of Colombia and Venezuela, this plant undergoes a dry season when it forms a noncarnivorous, short-leaved, succulent rosette. During the wet months it produces long,

narrow leaves 1/4 inch (6 mm) wide and around 6 inches (15 cm) long. The flowers are violet with purple veins. This species would be difficult to grow since it comes from such high elevations; during the wet season, night temperatures can drop to near freezing.

Pinguicula calyptrata
Native to the northern Andes, this species has oblong leaves in 2-inch (5 cm) rosettes with small lavender flowers.

Pinguicula involuta
From the southern Andes, this species grows in chilly, wet climates, with small rosettes 2 inches (5 cm) across and flowers that range in color from white to violet.

Pinguicula chilensis
Similar to the above, this is also an Andean species found in Chile and Argentina. The small leaves are oval, and the flowers are blue or white with blue veins.

Tropical Pings

Butterworts reach their height of diversity and beauty in Central America and the Caribbean. While forty years ago the number of *Pinguicula* species in the world was estimated to be around thirty, that number has tripled over the past few decades as intrepid botanists and collectors have scoured the hills and canyons of Mexico, searching for these plants in the most unlikely of places—and finding them. While the companions of carnivores in other parts of the world are commonly plants such as sphagnum mosses, cranberries, and bog orchids, it is not unusual to find butterworts in Mexico growing with agave, succulents, and *Tillandsia*. Instead of searching out wet, peaty soils to find butterworts, collectors look for dry cliffs of gypsum or moss-covered tree trunks. If you happen to be in Cuba or Haiti searching for pings, you might have better luck looking up instead of down, as they are sometimes found there on branches and twigs.

The incredible adaptability of *Pinguicula* in this region of the world is further shown by the fact that most species in the tropics are part-time carnivores. Mexico and the Caribbean are not equatorial places, but lie

Pinguicula moranensis.

The succulent winter leaves of *Pinguicula potosiensis.*

on the edge of the subtropical zone. While warm most of the year, the winters can be very dry, and the summers wet. What's a carnivore to do, with no water to help make all those glues and digestive juices? Turn into a tuber underground, like a winter-growing sundew? Tropical butterworts have a different strategy. When the rain stops falling, most turn into succulents. They give up their large, sticky summer leaves, quit the wholesale massacre of gnats and small flies, and transform into drought-tolerant rosettes of harmless dry leaves, inconspicuous and plain. Well, not quite, for it is during this otherwise boring succulent stage that most tropical pings do penance by sending up their colorful, glorious flowers. In this way, they are truly redeemed.

Mexico, where more butterworts grow than any other place in the world, is a diverse country. States like Oaxaca in the south are warmer and wetter than Tamaulipas in the northeast, where the Gulf Coast of Mexico meets Texas. While the warm temperate *P. pumila* is found in the wetlands of eastern Texas, traveling south along the Gulf Coast the genus does not reappear until about a hundred miles (160 km) south of the Texas-Mexico border. More rain falls in Mexico's coastal areas, while the interior mountains are not only drier, but cooler. Mexican butterworts can be found at high elevations where the dry winter nights can be surprisingly chilly, while the summers are warm and wet.

Botanists group *Pinguicula* species by the similarity of their flower structure, but this can be rather controversial. To further compound the problem, many butterworts that were described and named in the 1800s are now either lost in cultivation, grouped with other species, or have yet to be rediscovered in the wild. Some species, such as *P. moranensis*, have many different forms found in countless canyons and valleys that may or may not be species in themselves. The taxonomist's headache turns into the horticulturist's nightmare, as species' names change or are eliminated. Also, travelers to Mexico often return home with seed or plants (not realizing this is against Mexican law) and introduce them into cultivation, with data on the plants virtually nonexistent. As an example, over the years I have been given so many plants described as "*P. moranensis*," some with dubious site identification numbers (Site one #2, Site two #4), that I simply code them with a letter in hopes that someone will eventually tell me what it is I am growing! My first "*P. moranensis*" I lettered 'A', and at last count I was on letter 'O'. That's fifteen plants that may or may not be *P. moranensis*.

Pinguicula moranensis.

This confusion is slowly resolving itself with the growing interest in tropical butterworts. A few varieties are appearing in the general nursery trade as companions to Venus flytraps and purple pitcher plants (often with cultivation instructions certain to guarantee the death of the poor plant). Also, short-lived societies such as the International Pinguicula Study Group, based in England in the 1990s, worked toward the goal envisioned by some CP growers, such as Adrian Slack, who dream that one day these beautiful plants may achieve a status similar to that enjoyed by *Saintpaulia*, the African violet.

In the following pages I will avoid the nitty-gritty, technical floral descriptions of most of these plants and try to keep it simple, so as not to lose the beginner. Tropical pings are a fun and easy group of plants to grow.

Pinguicula moranensis

This is the most popular species to grow, in all of its forms and varieties. They make excellent windowsill, terrarium, and greenhouse plants, and can be in flower for many months of the year. The many different forms vary in leaf shape and size, but typically they have carnivorous summer rosettes up to 2 to 8 inches (5–20 cm) in diameter, the leaves oval to oblong, from pale green to suffused in red. The leaf margins are slightly incurved. In winter and early spring, the rosettes transform into succulents from 1 to 3 inches (2.5–7.5 cm) across. These noncarnivorous leaves are spoon to wedge shaped, tough and thick, and may number from a few to many dozens. *P. moranensis* are usually in flower twice a year. The flowers are usually pink, but there is also a white-flowered form, 'Alba'. Many blooms appear from the small succulent rosette almost continuously during winter and spring, some lasting for several weeks. When the summer leaves appear, the plants usually stop flowering for a couple of months. By late summer they experience a second flowering period. The summer leaves persist through autumn, but by early winter they die away and return to their succulent growth.

Pinguicula moranensis 'A'.

Of the many varieties I grow, I have never been able to make a positive match with similar forms grown in other countries. Of my own letter-coded plants, here are some of the best:

Pinguicula 'A'

Produces the largest summer rosettes, sometimes measuring more than 8 inches (20 cm) across, with leaves heavily mottled in red. The flowers are enormous—more than 2 inches (5 cm) in length—and pink with much white streaking toward the throat. The winter growth is a compact rosette of short, thick leaves. This variety occasionally clumps. Unfortunately, this is one of the few varieties whose succulent leaves mysteriously rot when used for propagation, so it remains rather rare and expensive when offered for sale.

Pinguicula 'D'

A handsome variety, with large, pale green, oval summer leaves and a tight winter rosette of thick, succulent leaves. It never clumps. The large flowers are pale pinkish lavender, the lobes almost rectangular.

Pinguicula moranensis 'G' in an abalone shell.

Pinguicula 'G'

The most popular form I grow; it can be in flower almost continuously, except during the transition times between summer and winter rosettes. The summer leaves are green and oval, and the winter rosette is short-lived, with a few spoon-shaped leaves 1 to 2 inches (2.5–5 cm) in length. This form regularly clumps. The beautiful flowers are deep pink with some white toward the throat. This variety is somewhat tolerant of wetter winters and rarely succumbs to rot.

The flower of *Pinguicula moranensis* 'G'.

Oaxacan *Pinguiculas*

Many forms of *P. moranensis* come from Oaxaca and are named after their location of discovery (or sometimes site numbers). A beautiful form called 'Site one #2' has summer rosettes up to 4 inches (10 cm) across, with densely compacted winter rosettes of numerous narrow and curved leaves. The striking flowers are narrow lobed and deeply purple pink, with a white "tongue" at the base of the middle lower petal. The two outer lower petals have a curious twist. A form called 'Mitla' has extremely narrow purplish lobes, described by Adrian Slack as spidery. 'Huahuapan' has lilac-colored narrow petals with darker streaks. 'Vera Cruz' has flowers of a deep rose color. In 'Site two #6', the smallish flowers are pink and rounded, the lower petals indented with white streaking toward their base. Elizabeth Salvia, wife of Barry Rice, registered *P.* 'Libelullita', which means "little dragonfly" in Spanish.

Pinguicula 'Libelullita', a cultivar named by Elizabeth Salvia. Related to *P. moranensis*, this may be a new species.

The flower of *Pinguicula agnata* 'True Blue'.

Pinguicula agnata

There are several forms of this species in cultivation, and some are so varied in appearance that they may warrant separate classification. None form true winter succulents, although their leaves may get smaller in size. In all *P. agnata* forms, the leaves are thick and fleshy and lack upturned margins. The true *P. agnata* from the Mexican state of Hidalgo has green, almost strap-shaped leaves that gradually press to the ground. The small flowers are pretty, with oblong petals of purplish blue, white at their bases, with a wide green throat and short spurs. A variety known as the 'Pale Flowered' form has petals of almost pure white, with only the slightest hint of purple around the yellowish throat. I grow a few other forms that have rounded, teardrop-shaped leaves that are often reddish in summer sun. In these, the flowers are almost round petaled, flat faced, and tinted purplish, with bright greenish yellow throats. One form has flowers that are occasionally scented like violets. It was introduced into cultivation by Leo Song of California State University in Fullerton.

Pinguicula gigantea

Originally registered by me as the cultivar *P.* 'Ayautla', this 1987 discovery by Alfred Lau is an easy-to-grow, exceptional species that is truly gigantic. The leaves can reach nearly 12 inches (30.5 cm) in length, although usually a bit shorter in cultivation. They are buttery green and arching, have no upturned margins, and incredibly have sticky glands on both sides of the leaf! The handsome flowers are tinted violet with a striking purplish edging. The plants were found in Oaxaca, growing on sheer cliffs in hot tropical sun, accompanied by *Tillandsias*. Like its cousin *P. agnata*, the leaves get smaller in winter but do not form a true succulent rosette.

Pinguicula gigantea, perhaps the largest Mexican butterwort yet discovered.

Pinguicula potosiensis

This plant has pale green summer leaves 3 or 4 inches (7.5–10 cm) long and slightly pointed, although some forms have attractive crimson leaves. The winter rosettes are densely compacted, with many dozens of short, narrow, succulent leaves. The medium-size flowers are purple, with slightly rounded lobes. This plant has been distributed as *P. moranensis* 'B'.

The diminutive summer rosettes of *Pinguicula esseriana* average 1 inch (2.5 cm) across.

Pinguicula esseriana

This is a popular miniature Mexican ping of recent discovery. It is from Tamaulipas and surprisingly tolerant of frost. The 1-inch (2.5 cm) rosettes are compact, with numerous short, thick, spatulate leaves with sharply upturned margins at their apex. In winter the leaves are similar but lose their margins and stickiness, being rather like the succulent leaves of a jade plant. The pretty, short-petaled flowers are lilac-pink and rather cup shaped, with long, downward-pointing spurs. This is a very tolerant species that can succeed well on windowsills; the winter leaves easily break free of the mother plant and root quickly, often forming clusters of plants.

Pinguicula ehlersiae
The rosettes of this species from San Luis Potosi are larger than those of *P. esseriana* and copper colored in strong light. The rather different flowers are flat faced, with rounded petals a lovely mauve.

The red carnivorous leaves emerging from the succulent winter leaves of *Pinguicula laueana.*

Pinguicula reticulata
This recent discovery, also from San Luis Potosi, is now famous among collectors for its beautiful flowers. They are cone shaped with short spurs and a hairy yellow throat; the nearly rounded petals are white with many violet veins. The leaves are rounded with rather long petioles, and there is little difference between summer and winter growth.

The brilliant flower of *Pinguicula laueana*, the only red bloom yet known among butterworts, although this species can also be rather variable.

Pinguicula laueana
Mr. Alfred Lau and his wife ran a boys' home on charitable donations in Veracruz, Mexico. He is well known for his many discoveries of Mexican flora, and among the finest was his 1978 discovery of this species, which entered cultivation in the early 1990s. From the Mixe Highlands of Oaxaca, *P. laueana* has summer rosettes of oval reddish leaves. In winter the succulent

growth is reduced to flat rosettes of smallish, overlapping leaves. But its glory is its flowers, shaped rather like *P. moranensis* but of a stunning, rich orange-red, the only red-flowered butterwort yet known. In some forms the flowers can be purplish, nearly orange, or fuchsia, and some have green foliage instead of red. A beautiful and easy species to enjoy.

Pinguicula gypsicola.

Pinguicula gypsicola

This strange Mexican ping is rather difficult to maintain in long-term cultivation, mostly due to its sensitivity to overwatering. It grows on gypsum cliffs that remain bone-dry most of the year, at which time it stays in a succulent growth of many dozens of tiny, densely compacted leaves nearly flat on the ground. But its brief summer foliage is most unusual. The carnivorous leaves are long and narrowly lance shaped, around 3 inches (7.5 cm) long by 1/10 inch (2 mm) wide. The flowers are rather similar to those of *P. moranensis,* with narrow, dark pink petals. This species must be kept dry during its succulent growth.

Pinguicula heterophylla

The summer leaves of this species are rather similar to those of *P. gypsicola*: thin, spidery, and arching. It is very odd in its having a long dormancy underground as an almost onionlike "bulb." It must be kept bone-dry during this rest period.

Pinguicula macrophylla has large carnivorous summer leaves but hibernates as an underground bulb during the dry winter.

Pinguicula macrophylla

I've had better luck with this species than with species like *P. heterophylla*; it also disappears to an underground, bulblike dormant bud for much of the year. The summer leaves, at the end of rather lengthy petioles, are large and oval. The summer flowers are deep purple with rounded lobes.

Pinguicula rotundiflora

Closely related to *P. reticulata*, this species was described in 1985. The summer leaves are rounded at the ends of long petioles, with upturned margins. The tiny winter leaves are also spoon shaped, but densely clustered. The flowers are cone shaped, with rounded pale violet petals and a darker throat.

Pinguicula colimensis

This butterwort is rare in cultivation. Many plants grown under this name are actually forms of *P. moranensis*. The true species grows on gypsum cliffs near Colima in Mexico. In winter the plant goes dormant underground as a small green "bulb" wrapped in dead leaves. The summer leaves are pale green and papery thin, without upturned margins. The flowers are large and beautiful, with wide, overlapping, dark pink petals, giving the flower a full, circular appearance. The long spur is thin and curved downward.

Pinguicula hemiepiphytica

From Oaxaca and described in 1991, this species has long been grown as a type of *P. moranensis*. It usually grows in moss on cliffs or tree trunks. The flowers are similar to those of a typical *P. moranensis*, but with more oval pink petals and a thicker spur. The oval summer leaves are pale green to copper. The distinctive winter rosette is a dense mound of small, pointed, spatulate-shaped leaves, often tinted bronze.

Pinguicula acuminata

This strange Mexican species has spoon-shaped leaves similar to those of *P. macrophylla*, and it also is reduced to an underground "bud" during winter. However, they are not related, and this species sends up winter flowers directly from underground. The stalks can be red, and the rounded petals are pure white, gradually turning pale lilac as they age.

Pinguicula cyclosecta

A striking species; the small spoon-shaped leaves of most forms are lavender to purple, as are the flowers.

Pinguicula emarginata
Another delightful small species with roundish leaves, the unusual flowers are small and, unlike other butterworts, have delicately ragged and jagged margins and are variably pale but highly veined in purple. Most unusual, pretty, and easy to grow. A clone I labeled as 'B' is more colorful than others.

Pinguicula cyclosecta.

There are a few tropical species that retain their carnivorous leaves year-round, growing in permanently damp conditions.

Pinguicula emarginata clone A x B.

Pinguicula lilacina
A widespread Mexican species; its oblong leaves have downturned margins. The pretty flowers are large and lilac colored with a darker throat. A yellow beard hints at its relationship to butterworts from the southeastern United States.

Pinguicula zecheri
This handsome species has arching, elliptical leaves. The frequent flowers are beautifully deep purple with a veiny white throat. Also from Mexico.

Pinguicula filifolia
This is an unusual tropical species from western Cuba, growing primarily on the Isle of Pines. My one attempt to grow it failed when winter arrived, probably due to cold temperatures. It grows in sand at the edge of freshwater lagoons, often accompanied by silver saw palmetto, a swamp-loving palm. As its name suggests, *P. filifolia* has narrow, threadlike leaves. Several color forms of the flowers exist: white, blue, purple, and lilac.

Pinguicula albida
This Cuban species is related to *P. filifolia* and *P. agnata* and maintains rosettes of oval green leaves year-round. The white petals are beautifully contrasted against the orange, starlike throat.

Pinguicula jackii

Another Cuban species, with 4-inch (10 cm) rosettes of oval leaves; the flowers are blue.

Pinguicula lignicola

This is an odd and rare species, also from western Cuba. A true epiphyte, it grows in trees and bushes attached to branches and twigs. The tiny rosettes of narrow leaves are barely 1 inch (2.5 cm) across. The large flowers are beautiful and white with orange centers, similar to *P. albida.* I expect it would be easy if grown as *Tillandsia* air plants are grown—on bark or branches, with frequent misting and overhead watering in a hothouse.

Pinguicula cladophila

This epiphytic species is similar to *P. lignicola*, but the rosettes can reach 2 1/2 inches (6 cm). It grows in mossy forest in Haiti.

An assortment of Mexican butterwort hybrids. Left to right, top row: *Pinguicula zecheri* x *laueana*, *P.* 'Weser', *P. laueana* x *agnata* 'True Blue'. Middle row: *P. emarginata* x *cyclosecta*, *P. agnata* x *gypsicola*, *P.* unidentified 'ANPA-C' x *laueana*. Bottom row: *P. agnata* 'True Blue' x *emarginata* 'B', *P. moranesis* 'A' x *laueana* 'Red', *P. agnata* 'True Blue' x (x 'weser').

Hybrids

While a few temperate butterworts occasionally produce natural hybrids, none is known among warm temperate and Mexican species. Among the latter this may be due to the isolation of most species amid Mexico's many mountains and canyons. Ironically, however, most Mexican species readily hybridize artificially in cultivation, and they often produce beautiful plants with increasingly showy flowers. The future of *Pinguicula* in horticulture, aside from further new introductions from the field, may lie in the production of ornamental hybrids—a development that has already created promising hints of what is to come. When one realizes the beauty and diversity of African violet

(*Saintpaulia*) hybrids, all resulting from two species with blue and purple flowers, the ping possibilities are mind-numbing. While many butterwort hybrids have been produced and sold, particularly from European nurseries that tissue culture the plants, only about two dozen have been officially registered with the ICPS thus far.

Pinguicula moranensis × *ehlersiae*

This beautiful cross results in plants with diamond-shaped summer leaves that turn coppery in the sun. It clumps prolifically, producing mounds of plants over time. The flowers have rounded, oblong petals of a rich bluish pink. Two cultivars have been named: *P.* 'Sethos' has a starry white throat; *P.* 'Weser' has darker veins and a white streak at the base of the lower central lobe. In winter the leaves get smaller.

Pinguicula moranensis x *ehlersiae*.

Pinguicula agnata × *gypsicola*

Leo Song first crossed these plants. The tissue-cultured clone is frequently sold in the mass market. The beautiful summer rosettes have arching, strap-shaped leaves that turn reddish in good light. The winter rosettes have many small blunt leaves, and during this time the plant must be kept rather dry or it is prone to rot. The pretty flowers are pale purple and funnel shaped, with light veining.

Pinguicula 'Gina'

Produced by Miloslav Studnicka of the Czech Republic, this lovely clone is the result of crossing *P. agnata* × *zecheri*. The *P. agnata*–like leaves have upturned margins. The flowers have pale violet, oval petals with dark purple margins and a deep purple mouth with yellow throat.

Pinguicula moranensis × *gypsicola*

Several variable clones of this cross have been named; all must be kept dry in winter. *P.* 'George Sargent' has lilac flowers, undulating strap-shaped leaves, and large winter rosettes of many small leaves. *P.* 'Hameln', 'Mitla', and 'Mola' have wider summer leaves.

***Pinguicula agnata* × (*moranensis* × *ehlersiae*)**

I grow a couple of attractive clones of this cross, produced by Leo Song. The leaves are short and rather oval, with full-petaled, pinkish blue flowers with darker veins.

Pinguicula rotundiflora esseriana

This cross produces nice, compact miniatures with small, pinkish-blue flowers, intermediate between the parents. The winter rosettes are small clusters of tiny, succulent leaves.

The flower of *P.* 'John Rizzi'.

***Pinguicula* × 'John Rizzi'**

In my California greenhouse in the summer, hummingbirds frequently make a beeline for our flowering Mexican butterworts, probably because the birds are familiar with pings from their winter sojourn south of the border. This sometimes results in seed, which I usually destroy, not knowing their pedigree. Once, however, I grew some plants from seed collected from a *P. moranensis* variety dubiously marked 'Superba'. Some of the resulting plants I sold before maturity, nicknamed "hummingbird mix." Our friend, the namesake of this cultivar, grew one of these plants to flowering size. Its exceptionally large, full blooms of deep pink petals warranted its preservation. The oval summer leaves are undulating and virtually marginless.

Pinguicula rotundiflora x *hemiepiphytica*.

Pinguicula rotundiflora* × *hemiepiphytica

This extremely attractive hybrid has summer rosettes $2^1/_2$ inches (6 cm) across with deep pink leaves and blood red margins. The nearly indescribable flowers are 3/4 inch (2 cm) across and a brilliant purple-fuchsia.

Pinguicula moctezumae x *zecheri*.

***Pinguicula* 'Pirouette'**

A registered cultivar with olive-pink leaves. The deeply folded margins make the summer leaves diamond shaped.

Pinguicula moctezumae* × *zecheri

This has strap-shaped green leaves that end in a point, with purple-pink ruffled flower petals.

Pinguicula laueana* × *cyclosecta

This features small roundish olive-purple summer leaves; the little flowers are lavender with darker purple veins.

Pinguicula agnata 'Red Leaf' x *emarginata* 'B'.

***Pinguicula agnata* 'Red Leaf' × *emarginata* 'B'**

The rosettes of oblong pink leaves can reach 2 to 3 inches (5–7.5 cm) across, and the curious flowers are pale with purple veins like those of *P. emarginata* but with rounded petals. Crossed by Damon Collingsworth.

CULTIVATION

(See Parts One and Two for further information.)

SOIL RECIPES: Temperate species: Use a mix of two parts peat, one part sand, and one part perlite. Warm temperate varieties do well in a soil of one part peat to one part sand. Mexican and tropical species have done best for me in equal parts sand, perlite, and lava rock and/or pumice, plus

continued

one part peat. Some growers add dolomite or gypsum to this; I have not found it necessary, although some gypsum helps with *P. gypsicola*. Some Mexican butterworts are also quite happy in looser, more airy soils. At California Carnivores we have some *P. moranensis, P. laueana,* and hybrids like *P.* 'Sethos' and 'Weser' thriving among our tropical *Nepenthes* pitcher plants in soils of long-fibered sphagnum, orchid bark, perlite, and tree-fern fiber. The winter succulent rosettes even tolerate the overhead watering the pitcher plants receive, probably due to good drainage.

CONTAINERS: Plastic or glazed ceramics with drainage holes suit most varieties. Warm temperate species can also do well in undrained containers, but you should let the water level fluctuate without drying out the soil. Mexican species do best in well-drained containers, but I have also grown them in shallow, undrained ceramics with very careful watering. I also enjoy growing Mexican pings in abalone shells (they appreciate the calcium) and chunks of lava rock with large nooks and crannies. I use the recommended soil for these, but I topdress the medium with a few strands of long-fibered sphagnum to keep it intact. Large-leaved varieties look best in wide, shallow containers. Butterworts have short roots, so pot depth is not important. Most do well in 3- to 4-inch (7.5–10 cm) pots. Larger-leaved Mexicans look best if leaves are supported in pots as wide as the plants.

WATERING: All temperate and warm temperate species should be grown permanently wet on the tray system, with frequent overhead watering. Use chilly water for your temperate pings. The Mexican varieties can be kept on the tray system with overhead watering while they have carnivorous foliage in summer and autumn. When the rosettes change to their small succulent forms in winter, keep the soil on the dry side, dampening it only slightly and occasionally. You can usually tell how dry a species enjoys its winter by the size of its succulent leaves. The tighter, smaller-leaved rosettes, such as *P. gypsicola* or the bulblike *P. heterophylla* and *P. macrophylla*, require bone-dry conditions. Species with larger winter leaves, like a few of the *P. moranensis* varieties or *P. agnata*, enjoy winter soils kept just slightly damp. Cuban species should be kept wet year-round, with only slight winter drying.

LIGHT: Most *Pinguicula* enjoy partly sunny locations or very bright light. Don't roast temperate pings in summer.

CLIMATE: As the name classifications suggest, temperate, warm temperate, and tropical butterworts can be grown outdoors in their appropriate climates. As a rule, the Mexican varieties withstand cooler, drier winters than those in equatorial climates and can survive winter-night lows in the upper 40s°F (8°C). The only Mexican butterwort I accidentally exposed to freezing temperatures was *P. esseriana*, which was unharmed after a dry freeze of several nights in the low 20s°F (-4°C). Summer temperatures for the Mexican species can be cooler at night, since most grow in the mountains and do well with nightly lows around 60°F (16°C) or a bit warmer.

DORMANCY: Temperate butterworts require chilly to frosty winters while they hibernate as dormant buds. Warm temperate species usually survive light winter frost but are best protected from severe cold, and they usually slow or stop growth in winter (but maintain their foliage). Mexican species that turn into succulents in winter are not truly dormant, but those that form "bulbs" are.

OUTDOORS: See the preceding Climate section. Mexican pings can be grown outdoors or on porches during the warmer months of the year, moving to windowsills for winter. The largest butterworts I ever grew were dinner-plate-size *P. moranensis*, grown under the dappled sun of redwood trees during the summer in coastal Northern California. In places like southern Florida, Mexican pings would thrive outdoors year-round.

BOG GARDENS: Temperate and warm temperate pings make interesting species for the outdoor bog garden in appropriate climates. They do best when lightly shaded by *Sarracenia* or ornamental bog grasses.

WINDOWSILLS: Temperate species are totally unsuitable as houseplants, but there are a few warm temperate butterworts, such as *P. lusitanica*, that can do well on a windowsill. Many of the Mexican butterworts make ideal plants for partly sunny windows and are charming for their pleasant winter and summer flowers. Those that excel are *P. moranensis, P. agnata, P. esseriana*, and *P. ehlersiae*, plus the vigorous hybrids and cultivars.

continued

Cultivation, *continued*

TERRARIUMS/GROW LIGHTS: Forget temperate varieties here. Some warm temperates—such as *P. lusitanica*, *P. caerulea*, and *P. primuliflora*—do nicely, but they do best with cooler winters. Most of the Mexican species thrive under grow lights, in a tank or not, at room temperature. Use the pot-and-saucer method to allow for drier winter conditions.

GREENHOUSES: Temperate species generally do best in cold houses, or they can be placed under benches during winter in cool houses and warm houses. Warm temperate species enjoy cool-house and warm-house conditions. Mexican species can survive cool houses, but thrive best in warm-house or hothouse conditions. Some Mexican species may grow in stove-house facilities, but none are truly equatorial. At California Carnivores we find that shading the greenhouse roof more heavily over our butterwort-growing areas suits them best.

FEEDING: Butterworts feed on small insects such as gnats and tiny springtails. You may feed your plants wingless fruit flies or small ants, or occasionally apply bits of dried insects to their leaves.

FERTILIZING: Fertilizing butterworts can be a tricky thing, as most are quite sensitive and can be damaged if fertilizer is applied to the roots. If you grow temperate or warm temperate species outdoors, such as in a bog garden, fertilizing is not necessary. The Mexican or tropical varieties can benefit from an orchid, an epiphytic, or a Maxsea fertilizer *sprayed only on the leaves*; avoid a heavy sprinkle, which allows too much to get into the soil. I have found acidic fertilizers to kill or damage some Mexican species, since many grow in alkaline soils. Warning: Fertilizers applied to the flowers will mar them with unsightly spots, so apply only to the leaves. One can also apply heavily diluted fertilizer to the leaves with tiny artists' paintbrushes.

TRANSPLANTING: Butterworts are not as fragile as they look, but they shouldn't have root disturbance during active growth. Transplant and divide gemmae of temperate species in late winter, or just as the new leaves appear in spring. Warm temperate species and Caribbean varieties likewise are best disturbed at this time. Mexican species can be transplanted and divided during the end of their succulent growth. Sometimes these plants lift themselves out of the soil with their ground-hugging summer leaves, surviving for months as their leaves protect their exposed short roots. When this occurs, you should break away the down-curved, older leaves and reinsert the roots in the soil. Most potted butterworts should have their soil changed every two to three years. A quick soak in SUPERthrive during transplanting will encourage new roots.

GROOMING: Usually old brown leaves simply dissolve, but they can be trimmed away.

PESTS AND DISEASES: These are few. Slugs and snails can attack pings, but they usually move on after a bite or two. Rarely do aphids attack the undersides of newly emerging leaves; if they do, use a recommended insecticide such as pyrethrin/canola oil. Flea collars in close proximity to the plants also control insect problems. In Europe, mites can attack Mexican pings, causing pale, deformed new leaves. Dicofol is an effective cure. Fungus can attack pings that are grown in dark, stuffy, overly humid terrariums and greenhouses, usually in winter. Apply a fungicide or, even better, change their environment.

PROPAGATION

Propagating butterworts is fairly easy to accomplish using a variety of methods.

SEED: Only *P. villosa* and *P. lusitanica* self-pollinate their own flowers. All other species can be pollinated by hand to produce seed set.

The flowers of butterworts are designed to be pollinated by "long-tongued" insects and animals such as butterflies and hummingbirds. The repeated in-and-out actions of their mouth parts, to get at the nectar deep in the spurs, effectively pollinates the plants.

When you closely examine the flower of a butterwort by peering down its "throat," you will notice a small, apronlike pad on the ceiling. This is the sticky female stigma. Hidden immediately behind this, and completely out of view, are the pollen anthers. To pollinate, use a toothpick or tiny paintbrush. Insert this past the stigma, and with a gentle upward swipe, withdraw. If you

continued

examine the pollinating object, you will usually notice a small amount of pollen grains on it. Carefully reinsert, dabbing the pollen onto the front-facing apron or skirt of the stigma. If you wish to cross-pollinate flowers (to produce hybrids or to cross several of the same species), dab the pollen onto the stigmas of the other blooms, being careful not to reach in past the stigmas of the blooms you are pollinating, since that could pick up pollen from the same plant and then self-pollinate the plant you are trying to cross-polinate.

After successful pollination the corolla, or petals, will wither and fall off within a few days. Over the next few weeks the small seedpod will swell, eventually turn brown, and split, revealing many seed. Collect immediately. Seed can be stored for a few months in the refrigerator. For best results, sow as soon as possible.

To germinate, sow the seed sparsely on the species' preferred soil medium. Keep damp, humid, and in bright light. Germination usually occurs within weeks.

GEMMAE: Temperate species that form winter resting buds are easily propagated with the large amount of small "baby buds" that are produced around the base of the "mother bud" during late winter. These gemmae are easy to remove with forceps just prior to the plant's spring growth. Place each cone-shaped gemma pointy-side-up on the plant's preferred soil mix. As spring approaches, they will send out roots and leaves and can be semi-mature by the end of the first season's growth.

LEAF CUTTINGS: All Mexican species that form succulent winter rosettes can be easily multiplied by this method. Use the small, dry leaves of the plant, just prior to or during the new growth of larger, carnivorous leaves. Using forceps, gently grasp each leaf to be removed without bruising it too much; with a soft tug the leaf will pull away from the rosette very easily. Up to half of the winter leaves can be removed without injury to the mother plant.

Lay these leaves right-side-up on the soil without burying them. Use their preferred soil mix as the medium. Keep slightly damp and humid; a propagating seed tray with a clear plastic dome or a pot covered with a plastic bag works well. Keep in medium-bright light. Budded plants and roots will rapidly appear at the base of the leaf. Promptly discard any leaves that rot.

The young plants can be potted and grown normally within a few months.

The Bladderworts

(Utricularia)

"I was forced to the conclusion that these little bladders are in truth like so many stomachs, digesting and assimilating animal food."

—Mrs. Mary Treat, 1875

The pinhead-size trap of a baldderwort.

Bladderworts are the strangest and probably the most highly developed plants in the world. Nothing about them is familiar or makes them akin to other flowering plants except their flowers and ability to photosynthesize. In fact, bladderworts are so weird that over the past two hundred years a parade of famous botanists has puzzled over not only the complexity of their truly amazing traps but also what to call their various body parts. Do they have leaves? Stems? Roots? These simple questions left scientists scratching their heads, but a study of the animal-catching, pinhead-size traps reduced the investigators to a cross-eyed stupor.

Bladderworts make up the largest genus of carnivorous plants and are the most widespread. They grow on nearly every continent of the world; they are missing only from the most frozen Arctic regions and most oceanic islands. At last count there were 228 species. Highly adaptable, bladderworts may be found in Alaskan swamps that are frozen most of the year; in quiet acidic ponds in sunny Florida; in wet, mossy South American trees; in fast-moving African streams; in seasonally wet Australian deserts; and even living in other plants, such as the bromeliads. Some bladderworts survive ice by turning into dormant, hairy buds called "turions" (specialized for overwintering). Others survive heat and drought by changing into underground tubers the size of a grain of rice. Others are annuals, dying off after a season's growth and coming back from seed.

In horticulture, *Utricularia*'s most popular attribute is its flowers. In fact, when the plant is not in flower, a neighbor's reaction to your pot of precious bladderwort might be "What an ugly pot of slime!" A month later, that same neighbor will exclaim, "What a lovely display of miniature orchids!" and beg for a cutting.

Bladderwort flowers may be as tiny as an ant or as large as a medium-size butterfly—and often as beautiful as the latter, but rarely as homely as the former. Most bladderworts—or "utrics," as hobbyists call them—have blooms somewhere from 1/4 inch to 2 inches (0.6–5 cm) in diameter. They are often truly orchidlike in appearance and offer a rainbow of colors: white, pink, purple, violet, yellow, and red—and often in multicolored combinations. Bladderworts are related to *Pinguicula*, and the flowers of bladderworts are somewhat similar in structure to those of the butterworts. They have spurs and two petal lobes, an upper and a lower, often in various shapes. The lower petal is usually the larger and more showy. Flowers may appear singly or in groups.

Typically the species grow in sunny, wet areas. Around 25 percent are true aquatics, free-floating in quiet ponds. The rest find their homes in permanently or seasonally wet or waterlogged sand, mud, or mosses, commonly in bogs and swamps or along lake margins. Several are epiphytes, growing on mossy trees, and at least one grows on barren wet rocks under waterfalls.

According to the most general description, utrics form creeping or floating stems that are usually thin and hairlike. They are completely rootless. Most of the plant is under ground or in water. The majority of the species produce leaflike appendages, called "photosynthetic stolons," that protrude along the soil surface. These may be 1/4 inch (6 mm) long or much larger—up to 5 inches (12.5 cm) and resembling true leaves. Aquatic varieties often have thin, branching leaves not much different from their stems. The entire plant may be only 5 inches (12.5 cm) across, while some aquatic varieties may grow several yards or meters long. When flowers appear, they grow up from the stems, protruding above the soil or water surface.

But it is the bladderlike trap that makes *Utricularia* a wonder of nature. Scattered along much of the plant's underground or aquatic stems and leaves are hundreds to thousands of tiny bladderlike traps. The traps are usually the size of a pinhead—or can be smaller than the period at the end of this sentence. In some species they are more substantial, perhaps 1/8 or 1/4 inch (3–6 mm) across.

These tiny traps are now known to catch small swimming prey in as little as ten- to fifteen-thousandths of one second!

The bladder traps of *Utricularia dichotoma.*

As early as 1797, an English botanist named James Sowerby noticed the bladders on an aquatic species and assumed them to be flotation devices. He also saw small insects in the bladders, but thought the creatures were just "lodging" there. During the mid-1800s, many botanists examined the plants but were often unaware of each other's findings until years later. Ferdinand Cohn found prey in the traps of dried herbarium specimens. In 1875, he put water fleas (daphnia) into an aquarium of live plants, and by the following day, all of the daphnia were inside the bladders. Charles Darwin thought insects forced their way into the traps. But it was the botanist Mary Treat, an American, who first observed that the prey were sucked into the traps through a small door in an instantaneous, vacuumlike manner.

The early twentieth century saw botanists slowly piecing together the complex nature of the bladderwort trap, and Francis Lloyd, in his 1942 *Carnivorous Plants*, explained that more than a dozen separate things occurred when a bladder caught prey—and most of them much faster than the blink of an eye.

The bladders are usually kidney, pear, or tubular shaped. They are attached to the plant by a small stalk. The traps are hollow and transparent, their walls only two cells thick. At one end of the trap is a small door, which can only open inwardly. When closed, an oozy mucilage around the door keeps it watertight. When set, the bladder's walls are concave, with a strong vacuum within.

Outside the bladder's trapdoor are several long, filamentlike hairs that usually form a funnel to guide prey toward the door. There are glands around the closed entranceway that may secrete a lure. Typical prey are microscopic organisms, such as paramecium and cyclops, and larger prey such as rotifers, water fleas, worms, and mosquito larvae. The biggest prey include newly hatched fish fry and tadpoles.

Minute trigger hairs grow at the trapdoor entrance. A mere touch of one of these hairs and—whoosh! The door swings inward, and the prey and any surrounding water are sucked inside by the trap's vacuum. The door slams shut. The prey is suddenly trapped inside its transparent vegetable prison.

Water is pumped out of the trap through the door within minutes. Again, the mucilage seal makes the bladder a watertight vacuum. As quickly as twenty minutes later the trap is reset and ready for its next meal. One bladder can catch more than a dozen prey.

Inside the bladder, glands secrete digestive juices that cover the prey. Within hours, the victim dissolves. These glands then absorb this nutrient soup.

Larger creatures caught by the traps suffer a particularly gruesome death. Tadpoles and mosquito larvae are often caught by their tails. They struggle helplessly to free themselves as they are slowly digested alive. As the trap resets itself, the prey's agonized thrashing will set off the trap again . . . and again . . . and again, until only the prey's head protrudes from the trap, too large to be sucked in through the door.

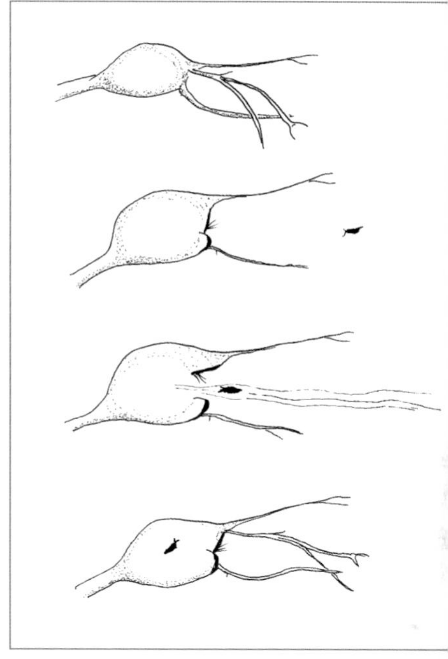

Upon touching the trigger hairs at the bladderwort's trapdoor, tiny prey are sucked inside faster than the blink of an eye.

In 1989, Peter Taylor of England's world-renowned Kew Gardens, in his impressive *Utricularia* monograph, reduced the number of recognized species from 250 to 214. There are now believed to be 228, give or take a few. Since bladderworts are such a large genus, and since most of the species are rare or unknown in cultivation, here I will review only some of the easier and more popular representative species. While curiosity about the traps is of interest primarily to those of a scientific bent, most hobbyists enjoy utrics for the flowers. Many species make fine windowsill and terrarium specimens, often putting on their flower show for months at a time. Others are beautiful in bog gardens, adding delicate color to the architectural boldness of plants such as *Sarracenia*. Still others thrive in greenhouses, their bright flowers rivaling orchids in beauty. Methods of growing utrics in ways that will highlight their tiny traps are outlined on page 283.

Terrestrial Bladderworts

These species are simple to grow and include some of the most attractive when in flower. They are native to all the world's climate zones, and some are pan climatic. Typically they grow in permanently wet, peaty sands that are sometimes flooded with shallow water, and they are frequent companions to other carnivorous plants. The small bladders feed on various swimming and crawling creatures that inhabit their waterlogged soils, such as fungus gnat larvae and tiny worms. Usually they produce carpets of short photosynthetic stolons along the soil surface, something like blades of grass pressed flat to the ground. Their flowers may be minute to substantial, held close to the soil or on stems 5 inches (12.5 cm) tall.

Utricularia subulata

Like *Drosera capensis*, this bladderwort has a reputation of becoming a weed in collections through its massive production of powdery seed. It is native to much of the world: Canada, south into South America, Africa, and Southeast Asia. They carpet the soil with fine, hairlike stolons or leaves barely 1/4 inch (6 mm) in length, and they appreciate occasional flooding. The multiple flowers appear in waves during warmer weather, on fine, purplish stalks that seem almost invisible. The bright flowers are sulfur yellow and about 1/4 inch (6 mm) across, with a large lower lip. The sight of masses of these flowers is particularly beautiful. Two things are peculiar about this species: first, sometimes the delicate flower stalks produce tiny drops of viscid glue along their length (probably to protect the blooms), something I discovered and wrote about in 1991. Second, the plants can also produce large quantities of seedpods without having flowered, a process known as "cleistogamous flowering." This species is excellent in bog gardens, terrariums, and greenhouses, and on windowsills.

Tiny drops of glue can be seen on some of the flower stalks of *Utricularia subulata*, a pleasant weed of carnivorous plant collections.

Utricularia livida

This very popular species comes from warm temperate to tropical climates in Africa and Mexico and is tolerant of light frost. The leafy stolons are short and blunt, often covering the soil surface. They flower continuously when the weather is warm and sunny. The lovely 1/3-inch (8 mm) flowers appear in rows of several along delicate 7-inch (18 cm) stalks. The large lower lip is apronlike, turning from white to violet in sun, with a yellowish streak at the spur's throat. They rarely produce seed, but spread rapidly through soil. A beautiful plant for the windowsill, terrarium, and greenhouse, and in bog gardens in suitable climates.

Utricularia livida, a wonderful free-flowering bladderwort.

Utricularia 'Merrie Heart'

This cultivar, which I named and registered, honors Mary Hart Lloyd, wife of Francis Lloyd. This Mexican form has particularly beautiful purplish flowers.

Utricularia 'Merrie Heart', a cultivar of *U. livida* from Mexico.

Utricularia sandersonii

From South Africa, this species is extremely popular due to its unusually pretty, rabbitlike flowers (which make the plant popular at Easter). Like *U. livida*, this species is suitable on windowsills and in terrariums and greenhouses, and in bog gardens in suitable climates. The short scapes hold up to half a dozen blooms, each about 1/2 inch (1.3 cm) across. The upper petal divides into two earlike lobes. The lower, skirted lobe

Miniature orchidlike flowers, with faces of angry little bunny rabbits, make *Utricularia sandersonii* a very popular plant. This one has grown in its tiny container for several years.

seems facelike due to its purple and yellow markings. The long curved spur bends forward under the petals. A form with wider, bluer flowers is unusual in producing threadlike runners, allowing the plant to spread into adjacent pots.

Utricularia bisquamata

A variable miniature from South Africa, this species flowers often and spreads from seed. The tiny flowers are beautifully multicolored in yellow, violet, orange, and white. There is also a form with larger flowers, as well as a tinier white form. It is also tolerant of light frost. This species is even more prolific at spreading from pot to pot in collections of CPs than the infamous *U. subulata*, and in fact colonized an entire outdoor bog garden several feet (2 m) across at California Carnivores in a matter of months!

The tiny but colorful blooms of *Utricularia bisquamata*.

Utricularia cornuta

This U.S. species grows from the Great Lakes into the southeast. The brilliant yellow flowers are similar to those of *U. subulata* but are much larger and are on tall, 12-inch (30.5 cm) stalks. It enjoys occasional flooding, with mass flowering as the water recedes. Excellent in the bog garden. *U. juncea* is also from the United States and almost identical, but smaller.

Utricularia resupinata

This species is found in parts of eastern Canada and the United States, as well as Cuba and Central America. It flowers after flooding recedes. The purplish and ruffled-looking flowers appear on short stalks and have a yellow throat.

Utricularia pubescens

This unusual tropical form is found in South America, Africa, and India. The peculiar soil-surface stolons look like tiny, dark green circular buttons. The shy flowers are large and lilac colored with yellow and white near the throat, on 10-inch (25 cm) stems. Best kept warm in terrariums, warm houses, and hothouses.

Utricularia dichotoma

A handsome species from Australia; its tall flowers look like lovely purple fans touched with yellow at the throat and usually appear in pairs. It appreciates waterlogged conditions and can survive periods of light frost.

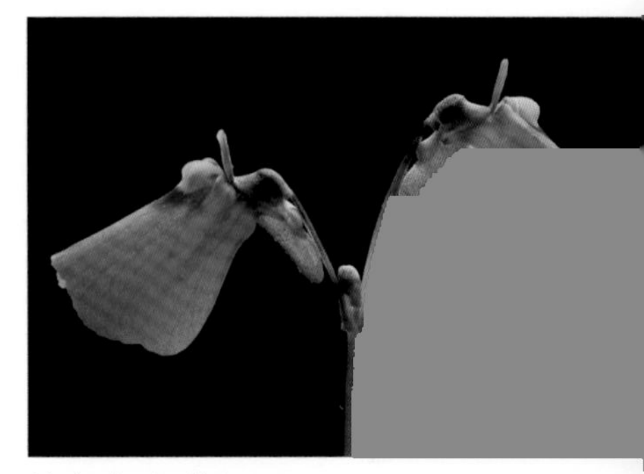

Utricularia dichotoma.

Utricularia praelonga

From Brazil, this species produces two types of stolons: one like thin blades of grass several inches (15 cm) long, the other circular and flat on the ground. The large yellow flowers appear on tall sprays up to 20 inches (50.5 cm) tall. Protect this plant from frost. It should be flooded occasionally.

Utricularia tricolor

This South American species has 1-inch-long (2.5 cm), kidney-shaped stolons held close to the soil. The lovely flowers are almost 1 inch (2.5 cm) across, pale violet with a ruffled skirt, and feature a puffed palate with yellow and white at the throat. This species also enjoys flooding and should be protected from frost.

Utricularia arenaria

This African species frequently produces tiny purple flowers touched with yellow. It spreads through seed and survives light frost.

Utricularia calycifida

This tropical form from South America does best in terrariums and hothouses. The unusual ground stolons look like teardrop-shaped leaves and are handsomely streaked in purple. These leaves can reach 2 inches (5 cm) in length.

Utricularia calycifida is an excellent terrarium plant.

The neat flowers, held close to the stem, are purple with yellow markings. Barry Rice has registered several forms of this species as cultivars, among them *U.* 'Yog-Sothoth', which has purple-veined leaves; *U.* 'Lavinia Whateley', an all-green plant with white flowers; and *U.* 'Cthulhu', also with veined leaves but with mauve rather than purple flowers. Mr. Rice adopted these cultivar names from hideous characters in the horror stories of H. P. Lovecraft. A warning: this species self-pollinates, so if you grow them near each other, seed of one cultivar may contaminate another!

Seasonal Bladderworts

Australia is home to a large number of unusual bladderworts that grow only during the wet season. In terms of climate, they may be considered tropical or Mediterranean. They survive through seed or tuber production. Most of these interesting species are only recently entering cultivation, usually through seed. For more information on their climates, refer to the woolly and tuberous sundew varieties (see pages 184 and 192).

Utricularia multifida

This Western Australian species is an annual, so flowers should be pollinated and seed collected. The large, beautiful flowers are pink, the lower petal divided into three rounded, incised lobes. This species was once called *Polypompholyx multifida*, but is now considered to be part of the *Utricularia* genus.

Utricularia menziesii

This lovely species from southwestern Australia is a cool winter grower, surviving hot dry summers by forming tiny tubers under the soil. It has a surface rosette of small green leaves. The magnificent flowers are bright red with an underhanging, pendulous spur. When this species is in full winter growth the soil should be kept very waterlogged.

Utricularia dunstaniae* and *Utricularia capilliflora

These two weird bladderworts are annuals from tropical, monsoonal northern Australia. Their tiny flowers are extremely peculiar—they look like insect heads. The flowers themselves are dull colored and minute, but the two upper petals look like long, thin insect antennae! It is believed that gnats pollinate the flowers.

Utricularia fulva
A beautiful northern Australian plant; its substantial flowers have seemingly ruffled petals of pale apricot yellow, with brownish red speckling at the palate. We have grown this striking species for several years in a heated tank at California Carnivores, and although described as an annual, returning from seed after the dry season, it continues to grow for us when kept wet and it flowers profusely. I suspect that similar species in this section may do likewise.

Utricularia fulva; the peachy-colored flowers last for weeks at a time.

Utricularia chrysantha
Thick and stocky flower stalks bear several pretty blooms, bright yellow with white lower lobes. From northern Australia.

Utricularia leptoplectra
This handsome species, also from northern Australia, has bluish violet flowers. The large lower lobe is deeply incised, giving the flowers the appropriate appearance of dangling boomerangs.

Utricularia lasiocaulis
Large, colorful flowers mark this northern Australian species. The upper and lower lobes, violet pink, are fanlike with a spot of yellow at the throat.

Tropical Epiphytic Bladderworts

These are all from the Caribbean, Central America, and South America, with many native to the tepui tabletop mountains of Venezuela. They typically grow in mossy pockets of leaf and bark debris in trees or on cliff sides, and they may die down to tubers during a drought. Most are popular greenhouse and terrarium plants. Some species may adapt to windowsills. Many are highland tropicals and are untroubled by nightly lows into the 40s°F (7°C), but for the most part they prefer warm houses and hothouses. Most produce large, showy, orchidlike flowers.

Utricularia reniformis

By far the most pleasing species to grow, this plant flowers in my warm house virtually nonstop from spring until fall. The leafy stolons are several inches (15 cm) tall on stiff petioles, the kidney-shaped blade 1 to 2 inches (2.5–5 cm) across. The flower spikes are more than 12 inches (30.5 cm) high, the several, long-lasting blooms 1½ inches (3.8 cm) across. They are pinkish violet, with a large lower skirt and puffed palate noticeably marked with two golden lines. The bladders are also large. The plant dies away each winter to underground tubers, at which time I keep the soil barely damp. This species should excel on the windowsill. To my amazement this species has survived in outdoor bog gardens at my nursery, the tubers surviving freezing temperatures and very wet conditions. Miloslav Studnicka of the Czech Republic registered this plant as *U.* 'Enfant Terrible'.

The large orchidlike flowers of *Utricularia cornigera*, more commonly grown as *U. reniformis*.

Utricularia cornigera

Originally believed to be a giant form of *U. reniformis*, and registered as 'Big Sister', this plant was described as a new species by botanist Miloslav Studnicka in 2009. The flowers are almost identical, on tall scapes and lasting for many weeks; its primary difference is the enormous leaves that can reach 4 to 8 inches (8–20 cm) across! When in flower this plant elicits gasps from onlookers at my nursery.

Utricularia longifolia.

Utricularia longifolia

A popular species with large, leafy, strap-shaped stolons 5 to 8 inches (12.5–20 cm) long. The bladders are large and often grow out of pots through the drainage holes. The large flowers are pink to violet, with a fanned skirt. The puffed palate has a prominent golden yellow streak. This species also has a dry winter rest period.

Utricularia alpina.

Utricularia alpina

This is a beautiful species when in flower. The leaf stolons are 5 inches (12.5 cm) long, paddle shaped, and pointy. The large flowers are creamy white with a yellow blotch on the palate. This also dies down to oval, opaque tubers in winter, when the soil should dry out.

Utricularia quelchii

A beautiful and rare species from the tepui Mt. Roraima (on the border of Venezuela and Guyana). The stolons are short, stiff, and teardrop shaped, with gorgeous purple-red flowers that are oval and skirtlike. A period of drier dormancy is necessary.

Utricularia jamesoniana.

Utricularia asplundii

Also from the tepui mountains. The stunning flowers are white with three lower, pointed lobes, and the palate is flushed violet with two prominent gold streaks.

Utricularia jamesoniana

Another rare species from the tepui mountains; its flowers have a purple, ruffled, skirted lower lobe with golden stripes on the palate. The incredible pendulous spur hangs below, thick and purple.

One of the largest flowers in the genus, that of *Utricularia humboldtii*.

Utricularia campbelliana

This species has small leafy stolons and large, bright red flowers.

Utricularia humboldtii

This is a strange species of utric that is commonly found in the water wells of large bromeliad plants growing on the tepui table mountains. It also grows terrestrially. Not only are the bladders very large—up to 1/4 inch (6 mm)—but the flowers are also the biggest in the genus at nearly 2 inches (5 cm) across. They are purple, with a large, undulating skirt and a golden white palate. The foliage stolons are erect, stiff, and teardrop shaped, but threadlike when young. The seed of this plant are bizarre, as the small green embryo is clearly visible in a transparent, flattened casing. The seed, when released, must immediately be sowed in water or they will dry out and die. Within twenty-four hours the seedling germinates as a tiny, star-shaped plantlet that grows rapidly. Surprisingly simple to cultivate, *U. humboldtii* succeeds in cool houses and warm houses. I like to grow them in bowls of waterlogged long-fibered sphagnum moss.

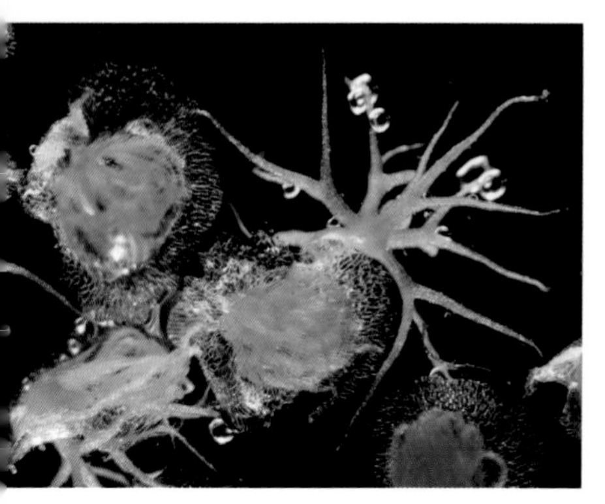
The bizarre seeds of *Utricularia humboldtii* germinating in water.

Aquatic Bladderworts

The free-floating, aquatic bladderworts are the species most familiar to the general public (through television nature programs) because they are the easiest of the bladderwort species to film. They are not necessarily the easiest to grow, primarily because popular varieties can reach enormous size and require wading-pool-size containers. Another hassle in cultivation is algae growth, which can inhibit utrics, and thus far there are no known treatments that kill algae without harming the bladderwort.

Daphnia help control algae while providing the plants with food. So do tadpoles—but they will eat utrics when they run out of algae.

Most aquatic utrics in cultivation are temperate species, and most form hairy turions during cold temperatures. They usually grow in still, shallow ponds of acidic water. Some of the plants flower en masse when the water table drops; a few form turions at times of drought. The foliage of the plants often alternates along the floating stems, producing whorls of threadlike leaves and other branches that produce the bladders. The often pretty flowers grow on stems sent above the water surface.

Utricularia gibba

If you wish to grow only one aquatic, it should be this species. It is by far the simplest to grow, surviving years in a container as small as a cup or bowl or even as an amphibious species in waterlogged peat. It also does well in the home or classroom on sunny windowsills or under grow lights. It even grows in the water trays of potted plants. The species grows in much of the world, in both temperate and tropical climates, usually in shallow water. It never goes dormant, and plants frozen solid return to growth when temperatures increase. The plants are small and fibrous, with 1/2-inch (1.3 cm) bright yellow flowers with a skirt and a puffed palate delicately penciled in red veins. The upper lobe forms an overhanging bonnet, and the spur looks like a curved tail.

Utricularia vulgaris and *Utricularia macrorhiza*

These two species are very large and very similar. The former grows in Europe, the latter from North America west into China. Their stems can exceed 10 feet (3.2 m) in length, and the bladders are large, up to 1/4 inch (6 mm). *Utricularia macrorhiza* often has bladders that change in color as they age, from green to red to black. The flowers are superficially similar to those of *U. gibba*, but larger. Their dormant turions are walnut size and hairy. This species likes a great deal of room—an appropriate container would be a children's wading pool. A pool of these

A damselfly rests on a flower of *Utricularia macrorhiza* in a Sierra Nevada pond.

plants with tadpoles and frogs; aquatic insects such as backstrokers, water beetles, and striders; daphnia; and water lilies or duckweed can provide an ecosystem that offers hours of mesmerizing fun.

Utricularia minor

A tiny species that grows in much of the northern hemisphere, this plant enjoys peaty slurries and has small, pale yellow flowers.

Utricularia purpurea.

Utricularia purpurea

This large plant grows from eastern Canada to Cuba and will go dormant only during freezes. It produces no foliage, only whorls of bladder traps. The flowers are beautiful: deep purple with a balloonlike palate.

Utricularia inflata

This unusual plant is common in the southeastern United States and has also been found, oddly, in Washington State. Similar to *U. macrorhiza*, it is strange because the flower stem is supported by a rosette of hollow flotation tubes, starlike in shape. The large yellow flowers grow upward from this. *Utricularia radiata* is similar but smaller. These plants go dormant only during freezing temperatures.

Utricularia volubilis

This species from Western Australia is an annual. It grows in winter ponds that dry out in summer. The plant is often anchored in sandy soil that is water covered, and it has bladders as large as 1/4 inch (6 mm). The lovely fan-shaped purple flowers are paired at the end of a scape that twines around reeds and grasses. Pollinate for seed. The seed can be sowed in a large bowl with wet sandy peat at the bottom. After germination, gradually add water over the course of a few weeks until it is about 5 inches (12.5 cm) deep. Provide bamboo skewers or thin branches for the flowering scapes to climb.

Utricularia reticulata

This tropical aquatic from Asia is often found in rice paddies. Its flower scapes also climb. They are a charming blue, with a paler inflated palate marked with dark lines.

CULTIVATION

(See Parts One and Two for further details.)

Most bladderworts are very easy to grow and propagate, and make nice companion plants to other CPs.

SOIL RECIPES: Aquatic varieties: One cup of peat well mixed into each gallon (3.8 L) of water. Epiphytic: A good mix is one part fine orchid bark, one part long-fibered sphagnum, one part peat, and one part perlite. Terrestrial: Use a mix of three parts peat to one part sand or perlite.

CONTAINERS: Plastic containers with drainage holes work best for terrestrial and epiphytic species. Most terrestrials also do well in undrained containers. If you wish to view the bladders on terrestrials, grow them in glass containers with removable black plastic sheeting or construction paper wrapped along the outside of the glass below the soil level. The traps will be visible when the dark covering is removed. Large aquatics require pools or tanks that hold a minimum of 50 gallons (189 L) of water. Smaller species succeed in containers that hold roughly 1 gallon (3.8 L) of water. Typical terrestrials do well for a couple of years in 3- to 4-inch (7.5–10 cm) pots; the largest of the tropicals may require at least 8- to 10-inch (25–30 cm) pots.

WATERING: Use the tray system for the terrestrials. Many appreciate periodic flooding and will do best in undrained bowls so the water level can rise and fall beyond the soil level. Epiphytic species enjoy a drained container (these can be set in shallow trays) that is watered from overhead, keeping the soil wet. If you're growing a species that goes dormant in winter, keep the soil only barely damp during that season. Aquatic species may need their peaty water changed if algae becomes severe. Gently rinse off the plants before introducing them to fresh water.

continued

Cultivation, *continued*

LIGHT: Sunny to partly sunny conditions for most species. Sun induces flowering.

CLIMATE: For outdoor growing, refer to the descriptions of individual species suitable to your climate zone.

DORMANCY: Most terrestrials just stop flowering during winter chills, but temperate ones should be kept colder. Even the larger tropicals often lose their leaves periodically but don't require dryness or cold and regrow after a couple of months.

OUTDOORS: Refer to the individual descriptions of species for suitable plants in your climate.

BOG GARDENS: An excellent place for terrestrials; they add color when in bloom. Larger bogs can include water-filled depressions or moats that allow you to grow smaller aquatic varieties.

WINDOWSILLS: Many terrestrials make fine sunny windowsill plants, particularly *U. livida* and *U. sandersonii. Utricularia gibba* is a good aquatic species to try. Of the epiphytics, *U. reniformis, U. longifolia*, and *U. humboldtii* may succeed.

TERRARIUMS: Warm temperate, subtropical, and tropical terrestrials do very well in lighted tanks, as do the tropical epiphytes. *Utricularia gibba* does nicely in a small glass jar in a potted tank.

GREENHOUSES: Refer to the species' climatic preference to see which species are appropriate for your greenhouse. Most terrestrials and epiphytics thrive in warm houses and cool houses. Large aquatics do best outdoors.

FEEDING: Aquatics can have daphnia (water fleas) and other microscopic life introduced to their water containers. Local natural ponds are a good source of these. The other varieties will feed on fungus gnat larvae and microscopic life that often grow on their own. If you flood your terrestrials, introduce daphnia.

FERTILIZING: Utrics appreciate light fertilizer about once monthly during their growing season. Fertilizers at one-quarter strength can be misted onto the foliage. Aquatics can have the water sprinkled similarly.

TRANSPLANTING: Most species are easily transplanted early in the growing season. New colonies of utrics should be started every two to three years.

GROOMING: No trimming is necessary except for the old dead flowers. Dead leaves on larger tropicals can be cut away.

PESTS AND DISEASES: Aphids can attack the photosynthetic stolons; use an appropriate insecticide or flea collar. Soil algae should be scraped off terrestrials. Algae can be problematic for aquatics, but daphnia help clear the water and feed the plants. Note that if algae are too sparse, tadpoles will eat utrics. Never use algaecides—change the water when infestation is great.

PROPAGATION

Propagating bladderworts is easy. Early in the growing season, simply remove sections of the plant and introduce these sections into fresh medium. That's it! With terrestrials, a section of soil, including surface stolons, of 1 or 2 square inches (2.5–5 cm) will do the trick. With epiphytes, likewise remove a section that includes the larger foliage stolons. Aquatics can simply have pieces separated, usually 3 inches (7.5 cm) long.

SEED: To produce seed, most need to be pollinated. Refer to *Pinguicula* pollination techniques (see page 265), as those flowers are similar. Insert toothpicks or (for tiny flowers) needles into the throat and spur, with an upward swipe, several times. Seed can be sowed on their preferred medium early in the growing season.

LEAF CUTTINGS: Leaf cuttings also work well with those species that produce leafy stolons. Simply pluck the stolons from the plant and treat as you would sundew leaf cuttings (see page 206). Larger leaves, such as those of *U. longifolia*, can be cut into smaller pieces about 1 to 2 inches (2.5–5 cm) long.

The Tropical Pitcher Plants

(Nepenthes)

"Can anyone see such marvelous things, knowing them to be only plants and feel no wonder?"

—*Gardeners' Chronicle*, 1849

A giant form of *Nepenthes rafflesiana*.

If there is a royalty among carnivorous plants, that distinction surely belongs to the *Nepenthes*.

Ever since their discovery by Europeans in the middle of the seventeenth century, tropical pitcher plants have inspired awe and wonder in anyone who has laid eyes on them. *Nepenthes* have a rich botanical and horticultural history, and the plants themselves are a virtual ecosystem of give and take in nature. The genus has the only species known to have devoured whole rats. And they are hauntingly beautiful, their pitcher traps often as elaborate and gaudy as any artistic creation of humankind.

While *Nepenthes* is primarily a plant of Southeast Asia, the first description of a species, *N. madagascariensis*, was given in 1658 by the then-governor of Madagascar, the Frenchman Étienne de Flacourt. He described, in a book on the history of the island, a strange plant with "a hollow flower or fruit resembling a small vase, with its own lid, a wonderful sight."

The second species described was *N. distillatoria* from Sri Lanka. When Carl Linnaeus first saw dried specimens of the plant, he was euphoric. He recalled Homer's *The Odyssey*, and the drug "nepenthe" that Helen of Troy threw into flasks of wine to alleviate soldiers' sorrow and grief. Linnaeus wrote, "If this is not Helen's *Nepenthes*, it certainly will be for all botanists. What botanist would not be filled with admiration if, after a long journey, he should find this wonderful plant. In his astonishment past ills would be forgotten when beholding this admirable work of the creator!" Thus in 1737 the genus received its Latin name. It is ironic that *N. distillatoria* is one of the simpler species of *Nepenthes*, compared with the elaborate ones that had yet to

Nepenthes khasiana.

be discovered, and that Linnaeus had no idea of the carnivorous nature of the plant, let alone any intoxicating influence the plant has on its prey. Like many others for years to come, he assumed the unusual pitcher leaves to be water-holding devices to help the plant survive drought.

It wasn't until the following century that *Nepenthes* had their heyday. Several things occurred to precipitate their rise in horticulture. One was imperialism, as Europeans began to explore and colonize Southeast Asia. In the 1700s orangeries were developed to grow the royal fruit citrus for kings, and soon thereafter glass greenhouses were built, enabling Europeans to grow exotic plants that were being discovered around the world. The Royal Botanic Gardens at Kew was started in England. In 1833 Nathaniel Ward invented the "wardian case," a sealed glass container that made it easy for exotic plants to survive long ocean voyages to England. In 1845 came the elimination of excise taxes on glass, resulting in cheaper and better greenhouses. Economies also boomed, so the middle and upper classes could afford such luxuries.

Nurseries also opened—for the first time plants were mass-produced for their ornamental value and sold to the public who could afford them. Among the first was the pioneering Loddiges Nursery in England, which introduced *N. khasiana* in 1825. James Veitch and Sons became the leader of such nurseries by the middle of the century. Hugh Low & Co. was another. These nurseries financed expeditions to faraway places such as Borneo, where exotic plants were collected and introduced into horticulture, *Nepenthes* being as sought after as palms, orchids, rhododendrons, and other ornamentals.

Also influential were the gardening magazines. Journals such as the *Gardeners' Chronicle* and *Curtis's Botanical Magazine* featured articles on the cultivation of *Nepenthes*, with beautiful illustrations and advertisements from suppliers.

By the late 1800s, *Nepenthes* were much in vogue. Most conservatory greenhouses on the estates of the wealthy boasted *Nepenthes* hanging from the rafters, tended by a gardening staff only the rich could afford. Fancy hybrids were winning silver and gold medals at flower shows. New species were being discovered and introduced.

After the turn of the nineteenth century, all of this came to an end. World wars, economic depression, fuel shortages—soon the dark early years of the twentieth century led to dark and empty greenhouses everywhere.

NEPENTHIANA

A study of tropical pitcher plants is virtually a who's who of early botany and horticulture. The following is a list of some of the personalities entwined among the vines of *Nepenthes*:

- Étienne de Flacourt: The French governor of Madagascar who first described seeing pitcher plants in 1658.
- George Everhard Rumpf: The famous tropical botanist known as Rumphius described *N. mirabilis* as 'Cantherifera' in a book written in the late 1600s.
- Carl Linné or Carl Linnaeus: The father of scientific nomenclature gave the genus the name *Nepenthes* in 1737.
- William Curtis: He started *Curtis's Botanical Magazine* in 1787; it is still in publication.
- Sir Joseph Banks: The great explorer and naturalist was involved with the early development of the Royal Botanic Gardens at Kew (Kew Gardens) and introduced *N. mirabilis* there in 1789.
- Father Joao Loureiro: A Portuguese priest in Vietnam, he described *N. mirabilis* as *Phyllamphora mirabilis* (marvelous urn-shaped leaf) in 1790.
- Sir Stamford Raffles: He was the founder of Singapore and started the Botanic Garden of Buitenzorg in Bogor, Indonesia. *N. rafflesiana* is named after him. Early 1800s.
- C. G. C. Reinwardt: He was the botanist of Raffles's garden. *N. reinwardtiana* commemorates him.
- Dr. William Jack: He discovered *N. rafflesiana* and *N. ampullaria* in Singapore around 1819. A surgeon for the East India Company, he befriended Sir Raffles when the latter was governor of Sumatra. He died tragically of "fever" at age twenty-five.
- Conrad Loddiges and his son, George: The first to introduce *N. khasiana* into cultivation (in 1825) through their Loddiges Nursery of Hackney, England. They were the first to make use of wardian cases to import *Nepenthes* and other exotics.
- P. W. Korthals: A Dutchman, he published the first monograph on *Nepenthes* in 1839, describing nine species.

continued

Nepenthiana, *continued*

- Joseph Paxton: He began publishing the *Gardeners' Chronicle* in 1841, which helped popularize *Nepenthes* in cultivation.
- Hugh Low: Son of the owner of Hugh Low & Co. nurseries in England, Hugh Jr. made three expeditions to Mt. Kinabalu in Borneo in the mid-1800s. He discovered four famous *Nepenthes*: *N. lowii, N. rajah, N. villosa,* and *N. edwardsiana*. He also introduced *N.* x *hookeriana* into cultivation.
- Sir Harry Veitch: Prominent member of the family that ran the Veitch Nurseries. *N. veitchii* is named after the dynasty, and Sir Harry is the namesake of *N.* x *harryana*. He introduced many species and hybrids of *Nepenthes* into cultivation. The Veitch Nurseries employed several of the most prolific *Nepenthes* hybridizers, among them Messrs. Dominy, Seden, Court, and Tivey. Many of their introductions survive today, and some bear their names.
- Thomas Lobb: An employee of Veitch Nurseries, he collected many new species of *Nepenthes*.
- John Dominy: Also an employee of Veitch Nurseries, he introduced in 1862 the first commercial hybrid, *N.* x *dominii*, plus many others.
- Sir Joseph Hooker: Son of Sir William, he became director of Kew Gardens in 1865. A friend of Darwin, he proved the carnivorous nature of *Nepenthes* and wrote the second monograph listing thirty-three species in 1873. *N.* x *hookeriana* is named for his father.
- Charles Curtis: Another Veitch employee and collector, he discovered *N. curtisii* (*N. maxima*), which was named for him.
- Marianne North: Famous botanical artist; *N. northiana* bears her name because Harry Veitch saw her painting of it and realized it was a new species. Today a gallery of her work remains on display at Kew Gardens.
- Frederick Burbidge: A collector for Veitch, he wrote the well-known *Gardens of the Sun* in 1880 (still in print). He discovered and named *N. burbidgeae* for his wife.
- Maxwell Masters: An editor of *Gardeners' Chronicle*; Veitch Nurseries named their beautiful hybrid *N.* x *mastersiana* for him and the work he did publicizing *Nepenthes*.

- James Taplin: An Englishman who moved to America, he produced many hybrid *Nepenthes* for George Such Nurseries in New Jersey in the late 1800s.
- J. M. Macfarlane: He wrote a revised monograph on *Nepenthes* in 1908, listing fifty-eight species.
- B. H. Danser: In 1928, he wrote a monograph on *Nepenthes* reducing the species count to forty-eight.
- Matthew Jebb and Martin Cheek: In 1997, these two botanists at Kew Gardens revised the genus, listing eighty-two species.

Beginning in the 1990s, the explosion of interest in carnivorous plants in general and *Nepenthes* in particular has not been matched since the nineteenth century. In fact, more new species have been discovered in the past couple of decades than at any other time period in the past. The species count is currently approaching 150! Many botanists, horticulturalists, and hobbyists have endured leeches, insect stings, injuries, exhaustive mountain climbing, and thirst, often in remote areas of civil unrest, exploring vast mountain ranges from the Philippines to New Guinea, in search of new *Nepenthes*—and they have found them. It is estimated by some that in the years to come the final count of species will be in the many hundreds. I will mention many of these individuals in the pages to follow.

Nepenthes are tropical pitcher plants that usually grow as climbing or scrambling vines. Most species are found in Southeast Asia, their center of distribution being the island of Borneo, but isolated populations are found as far from this center as northeastern India, Madagascar, the Cape York Peninsula in northern Australia, and New Caledonia. Besides Borneo, most species are found in Sumatra, the Philippines, the Malay Peninsula, Sulawasi, and Indochina.

Nepenthes are not typically jungle plants; they prefer more open and sunny ridges, slopes, meadows, fields, and stunted forests. Only 30 percent of the species are found in lowland areas where the days are hot and the nights warm. The majority of *Nepenthes* are highland or mountain plants, preferring warm days with cool nights. Humidity and rainfall are both high in the habitats supporting the plants.

Nepenthes are found in a variety of soils that are permanently wet throughout the year, although some survive droughts or brief drier seasons. The soils are kept moist by frequent rainfalls, foggy mists, or seeps and springs. Some *Nepenthes* are native to marshes or swamplands. The soil itself is often a shallow layer of leaf litter, decomposing bark and twigs, and mosses, including sphagnum. Although this loose and airy soil is generally acidic, it may overlie a foundation of ultrabasic rock like serpentine, or sandstone, or even alkaline limestone. Some *Nepenthes* grow epiphytically, their roots in mossy, leafy debris caught in the branches of trees. Sometimes the plants grow in wet sand or gravelly seeps. Common companion plants are ferns, grasses, shrubs, and stunted trees. As with most carnivorous plants, these habitats are low in nutrients, and moving water carries away what little minerals are in the soils.

The seed of *Nepenthes* are very thin and filiform, and so lightweight they can be carried off by the wind—their primary method of dispersal. Soon after germination, tiny rosetted plants are formed, and after one year's growth the entire plant may be a mere 1 inch (2.5 cm) in diameter, the tiny pitchers erect at the end of broad, leaflike petioles. As the plant grows year after year, the rosette spreads to a diameter of 3 inches (7.5 cm) in the smallest species to a few feet (1 m) in the largest. The extensive root system is very brittle and hairlike, usually from a thicker, sometimes branching taproot.

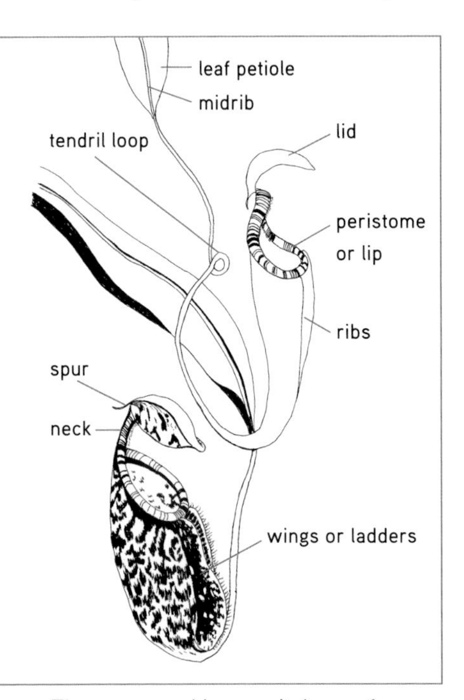

The upper and lower pitchers of *Nepenthes rafflesiana*.

The leaflike petioles are usually lance shaped but may be nearly oval or truncated, and they have a prominent midrib vein down the center. At the end of the leaf this midrib extends into a tendril, the tip of which is the immature pitcher. Not all leaves form pitchers, but in those that do, the tendril lengthens and the tip begins to grow and swell, ballooning into a hollow, sealed pitcher. When mature, the lid pops open, and the trap quickly reaches its full development. The lip or peristome unfolds, and the often bristly wings form "ladders" along the front. The pitcher quickly colors up in the sun, glistening with nectar droplets and often gaudy patterns to lure unsuspecting prey. These lower pitchers produced by the rosette sit on the ground facing outward. They are usually tubby and squat. When the pitchers open they already have digestion fluids in them, and they will secrete more once the plant starts to catch prey. It can take weeks to months for a pitcher to develop, and its lifetime may be similar.

A rosette of ground pitchers may take five to ten years to mature. Then the climbing stem begins to grow. The leaves of the climbing stem look similar to those of the rosette, but the tendrils and pitchers can be dramatically different. As the tendril elongates, it slowly moves about, groping through the air for something, like a branch or twig, to grab hold of. (This movement is best seen in time-lapse photography.) The tendril then forms a single loop, whether or not it finds anchor. But in order to form a hanging pitcher, support is usually needed. The tip of the tendril then swings upright like a hook, and it quickly swells into a pitcher. These upper or climbing pitchers can be so radically different in appearance from the lower ground pitchers they can seem to be of another species. They join the tendril from the rear rather than the side of the trap. They are usually more graceful and funnel shaped than the bulkier lowers, and they lack bristly wings along the front. In most species they are less colorful than the ground pitchers, and in fact can be pale green to yellow to nearly white. It's possible, as I have suggested in my *CPN* column, that pale-colored upper pitchers can be attractive to flying insects in moonlight.

Nepenthes 'Lady Pauline': upper pitchers on the left, lowers on the right.

Nepenthes produce climbing stems for the purpose of having their flowers higher in the air and sunshine than the surrounding vegetation. The climbing stems may be fairly short in some lower-growing *Nepenthes*, such as *N. ventricosa*. In others they may grow several feet to many yards (1–20 m) in length, scrambling over bushes or climbing into trees. After flowering, the stems usually continue to grow, and they may flower repeatedly for several years. Meanwhile, down below, new shoots appear at the base of the stem. These rapidly develop into large rosettes of new ground pitchers. A wonderful thing about *Nepenthes* is that usually every year a new rosette is formed that eventually becomes a climbing stem. Thus most species are a continuously rejuvenating mass of ground rosettes and pitchers with many climbing stems and hanging pitchers of various ages.

Male and female flowers are found on different plants, but look similar to each other. They are more odd than beautiful, but in some species can be colorful and attractive. The long flower stalks arise from the stem and are held upright. Each stalk has dozens to hundreds of small, densely packed blooms. Each bloom is a single flower, but occasionally they are joined in twos and threes. The individual flowers have short stalks and do not have true sepals or petals. Instead they have four short, teardrop-shaped tepals, and from this arises the small male anther or female stigma, depending on the sex of the plant. Stigmas are usually sticky and green, while the anthers are capped with a head of yellow pollen. Wind probably carries most pollen to female plants, but the tepals produce nectar to entice pollinators such as ants, beetles, and small flies. About 70 percent of plants in the wild are male and 30 percent female; thus males are also more common in cultivation. When pollinated, the female ovary swells, turns brown, and cracks open, releasing hundreds of fine, threadlike seed, the embryo a small nub in the center. One flower spike can produce thousands of seed.

Nepenthes flowers: male on the left, female on the right.

But it is the pitchers that make *Nepenthes* so famous. True leaves, the pitchers may be small and dainty to large and almost woody. They also have a fascinating life of their own, and apparently are more than just stomachs for the plant—they are a complete ecosystem of life and death.

The whole plant is covered with nectar glands that supply food for insects such as ants. Nectar is heavier along the tendril, and rather copiously produced by the pitcher, particularly along the ladderlike wings, around the liplike peristome, and under the lid. The lid does not move once it has opened (despite what is commonly supposed); rather, its function is to prevent rain from entering and diluting the contents too quickly. However, some species have small or narrow lids that freely allow rainwater into the trap; these act primarily as a nectar-baited lure.

Insects, primarily ants, visit the pitchers in great numbers. They are led by nectar and color patterns to the underside of the lid and the slippery peristome. For many insects, the nectar has an intoxicating effect. After feeding for a while, some insects can appear to be in a drunken stupor, walking or spinning in circles. Many of these lose their foothold, falling from the lid or peristome into the depths of the trap. It has recently been realized that peristomes wetted by rain become much more slippery, and that insects seeking refuge under the lid during rain can be catapulted into the pitcher by the force of raindrops hitting the lids of some species.

When a pitcher first opens, the secreted solution inside is fairly neutral in pH. But as insects are caught, their struggles apparently signal the pitcher to secrete acids and enzymes in large quantity. This liquid is often thick and almost syrupy, so the prey sink quickly and drown.

The interior of the trap is divided into two zones. The upper is the waxy zone, where most insects find it impossible to climb, their feet becoming clogged with a slippery substance. The lower digestive zone is covered with hundreds to thousands of large glands clearly visible to the naked eye. These glands secrete the juices that rapidly dissolve the soft parts of the prey. A fly can be digested in a couple of days. The glands then reabsorb nutrients from this soup. The carcass or exoskeleton sinks down to the growing graveyard of corpses at the bottom of the trap.

Strangely enough, tropical pitcher plants don't eat all of the insects and animals that visit their fanciful and dangerous traps. In fact, more than 150 creatures make the pitcher plants their home or otherwise have a mutually beneficial relationship with the plants during at least some part of their life.

An ant feeds on the nectar fangs of *Nepenthes bicalcarata.*

The simplest of these "friendships" can be found with ant colonies that make their nests near *Nepenthes*. While countless ants are caught and eaten by the plants, it has been found that at times of drought (when nectar is otherwise scarce), the pitcher plants sustain the ant colonies by offering sugary nectar for them to feed on. What effect the drugs in the nectar have on the ants is not yet known. It may be that only the nectar of the pitcher causes intoxication, and not from other parts of the plants.

Numerous mites and microscopic organisms, plus mosquito and fly larvae, live completely unharmed in the digestive juices of the plants, even when the acidity of the juice is as strong as 3.0 on the pH scale (the lower the pH, the more acid the substance). These creatures act as scavengers and may possibly help the pitchers with digestion.

A species of golden ant is known to drill holes into the thick, hollow tendrils of *N. bicalcarata*, where it raises its young. The adult ants feed on trapped prey. Drummer ants are well known on some species of *Nepenthes*. These solitary ants claim a plant as their own, and when threatened they beat their abdomens on the lids of the pitchers to scare off intruders. If drummer ants fall into the trap, they can easily and mysteriously escape.

The red crab spider is a common resident of *Nepenthes*, sometimes living in up to 35 percent of the plant's ground pitchers. It attaches itself to the interior of the trap by a small thread. It will swing on this and snatch flies that fall into the digestive juices, and has even been known to "fish" mosquito larvae out of the fluid. Amazingly, when threatened the red crab spider will plunge into the juices, only to haul itself out by its safety line when the threat has passed!

The pitchers of *Nepenthes* seem to go through stages of productivity as they age. Early in their life, they catch insects for the plant's benefit. But as they get older, their contents may become diluted with rain, or deteriorate and dry out. Many insects and other creatures then move in, feeding on the carcasses of the prey or making nests out of the once deadly traps—nature's recycling at its best.

Humans, too, have utilized *Nepenthes* for more than their beauty. Travelers have often used older pitchers filled with rainwater as a source of drinking water. As repulsive as it may sound, even insect-debris-laden water is refreshing to those suffering thirst in the tropics! The pitchers can also be cleaned out and used as water scoops.

Various medicinal uses have been beneficial to native inhabitants of Southeast Asia. The sterile solution in unopened pitchers has been used as an eyewash, an asthma reliever, and a painkiller during childbirth. (I once applied the fluid to a mild skin burn, and was amazed at the immediate relief.) The roots of *Nepenthes* have also been used to regulate menstruation and to help reduce fevers. Various parts of the plants have also been used for indigestion, heartburn, stomach ailments, and dysentery.

The climbing stems of *N. ampullaria* were once commonly used like rope to bind fences and other construction. Today, larger pitchers are still used as cooking tools: rice is often cooked inside the pitchers, as some believe the taste of the grain is enhanced by this method.

How have *Nepenthes* evolved? This has been debated since Darwin's time, and one fact stymied evolutionists for well over a century. Tropical pitcher plants could not have evolved like the pitcher plants of the New World. They are not descendants of ancestors that had "simple" rolled-up leaves joined at the seam, like a typical *Sarracenia* or *Heliamphora*. In fact, DNA and molecular biological studies have shown that *Nepenthes* share common ancestors with sticky-leaved carnivores like *Drosophyllum, Drosera*, and *Triphyophyllum* (to be discussed in the next chapter). It is believed that tropical pitcher plants evolved from some remote ancestor that had sticky, cup-shaped leaves that eventually transformed into pitchers.

The oldest fossil evidence of *Nepenthes* is of pollen found in Europe dating to the Eocene epoch, which ended around fifty-eight million years ago when our planet was quite warm. The most ancient relic species still in existence are survivors in Madagascar, the Seychelles, Sri Lanka, and northern India. These relic species are rather simple in design. Apparently the genus migrated to tropical Southeast Asia, where an explosion of new and more elaborate species evolved, most of them quickly, and perhaps within the last million years or so.

The Lowland Species

Species that grow below 3,000 feet (914 m) are considered lowland. They experience hot days, warm nights, and continuous high humidity.

Nepenthes gracilis

This fine and graceful scrambler is native to Borneo, Sumatra, Peninsular Malaysia, Singapore, and Sulawesi, and in some of these areas it is still a common roadside weed. The leaves are long and narrow, up to about 8 inches (20 cm). The small lower pitchers are 2 to 3 inches (5–7.5 cm) tall, cylindrical with a tubby base, with fine, eyelashlike wings and a thin, circular peristome and lid. The upper pitchers are similar but lack wings and can be twice as large. Several forms exist. The common form is green with many red spots along the upper half of the pitcher. Another beautiful form has pitchers of a full, deep red. Some forms are nearly blackish. Easy to grow and an excellent beginner's plant, it is perfect for the room-temperature terrarium and can have its fast-growing narrow vines pruned back severely to encourage bushier growth. Easy to root in water. This was my first *Nepenthes*, which I grew in a terrarium under grow lights.

Nepenthes rafflesiana

This magnificent species is extremely variable, with many forms native to Borneo, Sumatra, and Malaysia. A large grower, it is a showpiece in the hothouse, or, when young, in larger, warm terrariums. The leaves can be fairly broad and 1 to 2 feet (0.3–0.6 m) in length. Several forms have been named, but they are confused in horticulture. Different forms have been hybridized, giving birth to a wide variety of handsome offspring.

This species characteristically has bulky lower pitchers with a thick, striped peristome; fine, sharp teeth; pronounced wings; and a tall, spiny neck where the peristome joins the lid. The lid is often large and vaulted, with two prominent keels running lengthwise. Upper pitchers are usually as ornamental as lowers, but can be very elongated and funnel shaped.

The typical forms have pitchers 4 to 5 inches (10–12.5 cm) tall with pale green backgrounds that are very heavily splotched in reds or purples. Some that are fully red with light green speckling and greenish wings were named *N. rafflesiana* var. *nigropurpurea* by Masters in 1882. *N. rafflesiana* var. *nivea*, collected by Burbidge, has cream-colored pitchers with red speckling and white hairs on the stem, while in *N. rafflesiana* var. *nivea elongata* the pitchers are longer and more narrow. Jumaat Adam, in 1990, described *N. rafflesiana* var. *alata*, which has ornamental, frilly wings on the upper portion of the tendrils. This variety itself can be variable, with red-blotched pitchers to pitchers predominantly green, outlined with purple in the peristome and wings. The most remarkable forms of *N. rafflesiana* have huge purplish pitchers over 1 foot (0.3 m) in length, with tendrils that can be very long, dropping their heavy pitchers from above greenhouse benches all the way to the ground, a distance of nearly 5 feet (1.7 m).

Nepenthes ampullaria

This is another common, variable, and startling species from Borneo, Sumatra, the Malay Peninsula, and New Guinea. This species primarily produces numerous ground pitchers that are round and squat, resembling bird eggs. They are usually 1 to 3 inches (2.5–7.5 cm) high, but can be larger. The unique peristome sits at the top of the pitcher, circular and funnel shaped. The lid is narrow and strapped, deflected from the opening and offering no protection from rain, which the pitchers readily collect. Two prominent wings sit at the front of the tubby pitcher. The leaves of the climbing stem rarely produce pitchers, but when the stem is very tall, clusters of pitchers can suddenly appear along its length. This species is detritivorous, often catching leaf debris as well as insects, which decompose in the pitcher fluid, often with the help of mosquito larvae and other creatures.

A red form of *Nepenthes ampullaria*.

Many varieties exist. In one, the pitchers are entirely green. The most common has green pitchers liberally spotted in red. There are several striking red forms, among them *N.* 'Cantley's Red', which is scarlet with light green flecking. *N.* 'Harlequin' is a stunning cultivar with red pitchers flecked in dark purple and green.

Nepenthes mirabilis from Papua on the island of New Guinea. This new variety has been nicknamed 'John Holmes'.

N. ampullaria is very popular for warm terrariums, as the climbing stems can be easily pruned back, resulting in clusters of ground pitchers. The plants also pitcher nicely in shadier conditions.

Nepenthes mirabilis

This is the most widespread species of the genus, its many forms found from southern China to northern Australia, including Malaysia and the Philippines. In its typical form, the leaves are paper thin with slightly fringed margins. A small grower, it makes a nice terrarium plant. The upper and lower pitchers are usually similar: cylindrical with a bulbous bottom, round mouth and lid, flattened peristome, and colored green to suffused with red. An interesting form called *N. mirabilis* var. *echinostoma* has a marvelously wide, oversize peristome that is flat and striped. *N. mirabilis* var. *globosa* has beautiful, tubby lower pitchers, deep red with flamboyant wings. It was widely circulated as *N.* 'Viking' to collectors after being poached by two local families from an unknown island off of southern Thailand. The families perished and the plants were rendered extinct in the 2005 tsunami, but other populations of var. *globosa* have been found on the mainland.

As an extremely variable species, *N. mirabilis* can be found from acidic peat swamps to alkaline limestone cliffs and has even been found to colonize coastal brackish marshes and seasonally dry tropical forests, surviving by thick, tuberlike taproots.

Nepenthes bicalcarata

This amazing plant is famous for its sharp, saber-toothed fangs that hang from the rear of its lid, making it appear rather dangerous. Native to Borneo, it prefers shaded peat swamps, and it can grow to enormous dimensions. The stem is thick, with long, broad leaves 2 feet (0.6 m) in length. The lower pitchers average 6 inches (15 cm) tall; they are round and squat and almost woody in texture, with prominent wings.

They are green to coppery orange or reddish. The broad peristome joins to form a tall neck capped with a large lid. The two hard, sharp fangs are an outgrowth of the peristome and overhang the pitcher's mouth. Nectar sometimes drips from these, giving the appearance of oozing venom. Ants find the fangs difficult to negotiate, and they often fall from them into the digestive pool below. The upper pitchers are similar, but lack wings and are yellow-green.

The lower pitcher of the vicious-looking *Nepenthes bicalcarata,* its sharp fangs overhanging its mouth. The frightened *Tillandsia* is *T. butzii.*

N. bicalcarata is easy to grow but needs hot, humid conditions. When young it succeeds well in terrariums, but it does best in roomy hothouses and stove houses. Long a favorite with collectors, it is another showy, fantastic plant that never ceases to amaze people who see it.

Nepenthes albomarginata

A small-growing scrambler from Borneo, Sumatra, and the Malay Peninsula, it has cylindrical pitchers around 6 inches (15 cm) long, with an oval mouth and lid, narrow peristome, and reduced wings. The beauty of this plant is the prominent white ring below the peristome, which appears almost hand-painted. Most forms have grayish green pitchers, but a lovely Malaysian form has red lower pitchers. An amazing recent observation among wild populations is that the white ring is covered in minute hairs that termites find rather tasty, resulting in a gluttonous feast for the plants!

A green form of *Nepenthes albomarginata.* The mystery of the white bands beneath the lip has only recently been solved.

Nepenthes reinwardtiana

This tall scrambler also comes from Borneo, Sumatra, and the Malay Peninsula and sometimes grows epiphytically in trees. The smooth, curvaceous pitchers have negligible wings and are long and thin, with a slight waist. The slanted mouth is oval with a thin peristome and oval lid. Its hallmark is two curious, waxy "eye spots" that usually appear on the upper interior back wall of the pitcher, which may be a lure for prey. There is a common green form and a more striking red form.

Nepenthes truncata

A spectacular plant from the Philippines, this large, coarse species has unusual leaves that are squared or truncated at their ends. The enormous pitchers have a smooth green exterior and a colorful interior heavily mottled in reds, pinks, and purples. The lower pitchers are fat and cylindrical with prominent wings. The slanted mouth is large, with a colorful, wide peristome that may be fluted along its edge and striped or golden orange. The lid is domed and held horizontally. The pitchers may reach 14 inches (35.5 cm) in length. Highland forms enjoying cooler temperatures have been introduced into cultivation, with spectacular giant pitchers often suffused with bronzy coloration.

Nepenthes northiana, a striking lowland species with enormous and beautiful pitchers.

Nepenthes northiana

This showy species was made famous by the nineteenth-century botanical artist Marianne North's colorful painting. It grows on limestone cliffs in Sarawak. The giant lower pitchers reach 14 inches (35.5 cm) in height and are bronzy green, heavily splotched with red. The large, slanted mouth has a huge fluted peristome, dark red with purplish stripes. The lower pitchers are fat and squat, whereas the similar uppers are more cornucopia shaped. I have had limited success with this amazing species, trying both alkaline soils with limestone added and a more typical sphagnum medium, but losing the plants due, I believe, to cooler winter temperatures in my hothouse.

Nepenthes merrilliana

From the Philippines, this is another huge species that does best in stovehouse conditions. The heavy lower pitchers can be over 12 inches (30.5 cm) in length, with a gaping mouth and prominent wings. The form I grow is mostly green with red spotting and curiously flat-bottomed pitchers, but other forms can be green with red peristomes or a startling red body with green peristomes.

Nepenthes veitchii

This beautiful species from Borneo is found from sea level to about 4,800 feet (1,463 m), so some forms can be considered highland plants as well. The squat pitchers are green with strong wings. The leaves and pitchers are hairy. Its main attribute is its tremendously flared, nearly vertical peristome, reminiscent of the gills of a fish. This predominant feature can be green to golden brown or, in some spectacular forms, beautifully striped in red. The oval lid seems to hang precariously at the top. This species can be a "tree hugger," the vines growing up tree trunks with the tendrils and pitchers wrapped around them. In highland areas the vines can trail along the ground, supported by the sitting pitchers. This species honors James Veitch, reknowned Victorian nurseryman.

The incredible tree-hugging *Nepenthes veitchii*.

Nepenthes clipeata

On the verge of extinction, this spectacular species has very few plants remaining, its numbers decimated by overcollection and forest fires. It grows only on one cliff side on Mt. Kelam in Borneo but has entered cultivation. The leaves are oval, and the tendrils bearing the pitchers come not from the leaf tip but from its underside. The purplish pitchers are flask shaped and can reach a foot (0.3 m) in height. The peristome is striped, and the unusual lid is a domed canopy over the oval mouth. The beauty of the plant is the way in which the pitchers seem to hover in the air beside the leaves, their tendril attachments hidden from view.

Nepenthes campanulata, once thought extinct, but rediscovered by Ch'ien Lee.

Nepenthes campanulata

The species name of this plant means "bell shaped" and the small pitchers are like upside-down bells, dainty and green. In this true miniature, there are no upper or lower pitchers; they all appear similar, and the pitchers reach no more than 4 inches (10 cm) high but are usually half that height, with stems no longer than 20 inches (50.5 cm). Most unusual, its runnerlike stolons can form leafy clumps of the plants. The pitchers are paper thin, with an equally narrow peristome, and the oblong lid is never more than half the size of the circular mouth. Discovered in Borneo on a single cliff face in 1957, they were all destroyed in a forest fire in the 1980s. In 1997, naturalist Ch'ien Lee, an American who moved to Borneo, discovered another population 250 miles (400 km) away! A marvelous plant for the warm terrarium.

Nepenthes madagascariensis and *N. masoalensis*

These two species are relic plants and the only ones found on the island of Madagascar. I never succeeded with the former species (the first tropical pitcher plant ever discovered by Europeans in the 1650s) because I was misinformed that it was a cool-growing highland variety and it promptly died. Both species have tubby red lower traps with narrow peristomes up to 5 inches (12.5 cm) tall. Upper pitchers are green, sometimes with red lids.

The Highland Species

Seventy percent of *Nepenthes* are tropical highland or mountain plants, growing at elevations of 3,000 to 10,000 feet (914–3,048 m) above sea level. Above the lowland heat of the rain forest, the mountain climate can be considerably cooler and wetter, especially at night. In the center of this zone, day temperatures average in the 70s°F (24°C) and by early morning drop into the 60s°F and 50s°F (16–10°C) or even cooler, but frost never occurs except at the very highest level. The mountains are often shrouded in cloud cover, the nights frequently misty or rainy. As a result, the ground, rocks, and stunted trees are typically covered with thick growths of sphagnum and other mosses. These elevations are known as elfin or mossy forests, and when the sun breaks through the thick clouds it is usually brief and in the breezy afternoons. In his legendary May 1964 *National Geographic* article, which greatly influenced the modern resurgence of cultivating *Nepenthes*, Paul Zahl described Mt. Kinabalu in Borneo, at 6,500 feet (1,981 m), as "eerie," "bewitched," and "sharply cool." This is the *Nepenthes* zone.

Nepenthes ventricosa

From the Philippines, this wonderful species is the first *Nepenthes* I usually recommend for beginners. It is a low grower, with compact leaves growing along the gradually scrambling, branching stems. The lovely pitchers are tubby and rounded, with a constricted waist and no wings (although baby plants may produce rudimentary wings briefly). The mouth is wide and oval, with a beautiful scalloped pink to red peristome, thick, tightly ridged, and sharp toothed. Variable, the best forms have lower pitchers up to 5 inches (12.5 cm), suffused in carmine with crimson blotches or entirely red. Upper pitchers are pale greenish yellow and smaller. The flowers can also be handsome and colorful. An excellent plant for the windowsill and terrarium.

Nepenthes khasiana

The first species introduced into cultivation in 1825, *N. khasiana* is an endangered plant from the Khasi Highlands of Assam in northeastern India, the only *Nepenthes* native to that country. The lower pitchers aren't too exciting, lacking much color, but the uppers are cylindrical and up to 8 inches (20 cm) long, tinted bronze with a reddish band below the peristome, and the underside of the lid can be quite crimson. It is tolerant of rather chilly night temperatures down to 40°F (4°C), and I once had a large specimen return from the thick stems after a brief freeze. I also grew a large plant at a partly sunny living room window for a few years. It is one of the relic species and nearly impossible to propagate from cuttings; however, tissue-cultured plants are available.

Nepenthes alata

A common and widespread species from the Philippines to Malaysia, this species is extremely variable and also grows in the lowlands. Many forms exist in cultivation. The common one has slim pitchers with a slight waist and a bulbous bottom; the lower pitchers have fringed wings. The peristome is thin, with an oval mouth and lid, but overall the pitchers are bland, with only a slight flush of pink. In a plant I grow called 'Highland Form', the pitchers are rather similar but more flushed with red. Other varieties may have green pitchers with attractive red peristomes.

By far the best variety of *N. alata* is *N. alata* var. 'Spotted Form' from Luzon in the Philippines. This beautiful plant is very adaptable to warmer or cooler conditions, often excelling on windowsills or outdoors in subtropical climates. The lower pitchers are plumper than the typical form, with strong wings, and heavily streaked and splotched with red, sometimes entirely cherry red with darker spots. This contrasts well with the pale interior of the trap. Upper pitchers are more funnel shaped but also nicely colored. The peristome is thin.

Nepenthes maxima

This rather gorgeous pitcher plant is widespread and variable, growing primarily in Sulawesi and New Guinea. Mostly from the highlands, where in New Guinea some forms can experience frost, it can also be found in lowland areas. Typically the lower pitchers are 6 to 8 inches (15–20 cm) tall but can be larger. They are heavily blotched and streaked in crimson on a pale green to olive background. Wings on the lower pitchers are prominent. The mouth is oblique, with an enormous, fanciful peristome that can be widely flared and fluted, colored pink to wine red. The interior pitcher is pale with reddish spots. The lids, often held upright, are oval to triangular and green suffused with reddish streaks and spots. Most curious are the hooked boss at the underbase of the lid and a thin, filamentous "tooth" hanging from the tip. Usually the upper pitchers are smaller, extremely funnel shaped, and lacking much color. There are also miniature forms of *N. maxima* with pitchers not much taller than 2 inches (5 cm). I grew a mystery plant for years until *Nepenthes* expert Ch'ien Lee visited and identified it as one of the diminutive forms. A fine plant for terrariums and windowsills, this is one of the most popular species and very easy to grow.

Nepenthes fusca

Another beautiful plant, this species from Borneo is also quite variable. The lower pitchers are narrow and up to 8 inches (20 cm) long. One form, which I call "Coppermouth," has a coppery orange peristome while the pitcher is blotched and spotted in brownish red. The more popular variety has long, cylindrical pitchers heavily marked in purplish red, with a stunning peristome so purple maroon it looks almost black. The unusual lids are very narrow and triangular. Upper pitchers are short and very funnel shaped, sometimes spotted, with unusually narrow and downward-curved lids.

Nepenthes fusca's lower pitchers. This form comes from Mt. Kinabalu.

Nepenthes stenophylla*, *N. faizaliana*, and *N. boschiana

These three species, also from Borneo, are related to *N. fusca* and share similarities, namely elongated pitchers—particularly the upper ones. Both lower and upper traps can be quite colorful, with purplish red streaking and dark peristomes.

Nepenthes sanguinea

This vigorous, fast-growing Malaysian species has large, plump pitchers up to a foot (0.3 m) tall, a big oval mouth with a large, upturned lid,

Nepenthes sanguinea, an easy and fast-growing highland species. This is a lower pitcher of the large 'Red Form'.

a medium peristome, and prominent wings. The magnificent 'Red Form' has lower pitchers fully scarlet on their exteriors, with pale to spotted interiors and a cherry-red peristome. The upper pitchers are funnel shaped and plump, pale green, with red spots along the upper portion of the trap and a striped peristome. Other forms of *N. sanguinea* have uniformly yellow-green pitchers, sometimes with red stems; still other forms can be chocolatey brown.

Nepenthes macfarlanei

From the Malay Peninsula, this variable species has heavy, fat pitchers with large mouths that are oval to teardrop shaped, thick peristomes red to purple, and medium wings. The lids are large and oval. The lower pitchers can be pale brown to reddish tan, with irregular reddish to purple spots. Upper pitchers can also be large, usually a beautiful creamy yellow with interior red spots, heavily striped peristome, and red-blotched, large oval lids. The underside of the lids are covered with fine, short hairs, unique among the genus.

Nepenthes gracillima, N. alba, and *N. ramispina*

These three species are also from the Malay Peninsula and have been confused in the past. *N. gracillima* is notable for its elongated upper pitchers nearly blackish purple, flecked with green. The miniature *N. alba* has upper pitchers almost pure white with light red flecks. *N. ramispina* has long, narrow pitchers, the lower ones nearly black, the uppers a very dark olive, and angularly shaped, appearing flat fronted.

Nepenthes tentaculata

Another small grower excellent for terrariums, this variable species comes from Borneo and Sulawesi. The flat front of the pitchers has handsome, bristly wings; the slanted mouth is almost triangular with thin lips. Most, but not all, lids have tentaclelike hairs on top. The pitchers are usually 4 to 6 inches (10–15 cm) long and may be green, spotted, or red. Much larger forms of this variable species are known with pitchers almost a foot (0.3 m) high. My favorite of the ones I grow have red lower traps with triangular green peristomes, the pitchers never taller than 3 inches (7.5 cm).

Nepenthes tobaica

This comes from Lake Toba in Sumatra; I grow two forms with pitchers no more than 3 inches (7.5 cm) tall, although some forms may be twice as large. My favorite has brownish red lower pitchers, with uppers green with red spots. The other is entirely green and rather plain. The peristomes are thin, the lids circular. These make great terrarium or windowsill plants since the scrambling vines can be cut back when lower shoots appear. Many color forms appear in the wild; some, like *N. reinwardtiana*, have eye spots.

Nepenthes mikei

This very eye-catching species, also from Sumatra, has small narrow pitchers, almost entirely blackish purple, which contrasts well with its pale, whitish green interior. A stunning plant for the terrarium.

Nepenthes mikei.

Nepenthes bongso and *Nepenthes ovata*

These two Sumatran species are related and physically similar, with bulky lower pitchers to about 8 inches (20 cm) high, and are notable for their wide, flared peristomes. *N. bongso*'s lower pitchers are entirely purple; it was long grown by me as *N. carunculata*, a now obsolete name. In *N. ovata*, the pitchers are green with a bright red peristome. Upper pitchers of both are cornucopia shaped, mostly green, with reddish or striped peristomes.

The giant lower traps of *Nepenthes spathulata.*

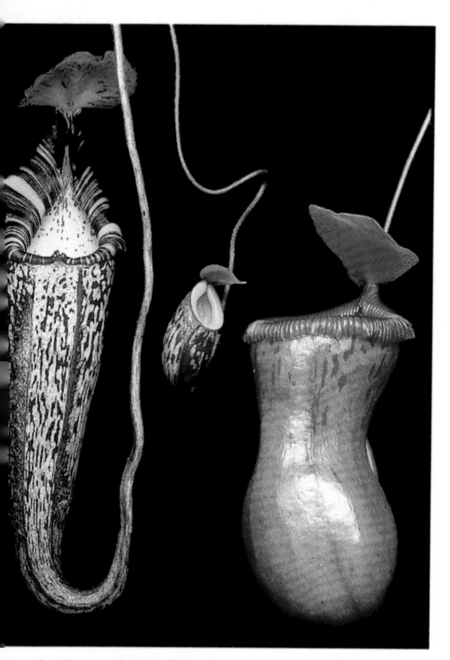

Left to right: The lower pitchers of *Nepenthes spectabilis, N. glabrata,* and *N. ventricosa.*

Nepenthes singalana

Also from Sumatra; its peristome is awesome—wide and deeply ridged, with sharp needlelike teeth. The lower pitchers are purplish red with strong bristly wings. The uppers are green with red mottling and surprisingly narrow, simple peristomes.

Nepenthes spathulata

Yet another Sumatran species, with a huge wine-red peristome and bulky green lower pitchers, the upper pitchers mostly green and less attractive.

Nepenthes spectabilis

This Sumatran species has beautiful pitchers, plump and cylindrical, with a creamy yellow background, purplish blotches, and a green-and-red-striped peristome.

Nepenthes eymae

I grow three forms of this species from Sulawesi. The lower pitchers are brownish orange and squat, with a substantial scalloped yellow to coppery peristome, and can be 3 to 10 inches (7.5–25 cm) tall depending on the clonal variety, with triangular lids. The upper pitchers are shaped like a wine glass, with an oval mouth and a very narrow lid, between 2 and 6 inches (5–15 cm), again depending on the clone. The pitcher rapidly constricts to the tendril, its interior sticky like flypaper, the digestive juices viscid like syrup. The whole plant is usually covered in fine fuzzy hairs.

Nepenthes muluensis

From Mt. Mulu in Borneo, this charming small grower has 5-inch (12.5 cm) cylindrical traps heavily blotched with purple on a creamy yellow background. In appealing contrast, the oval peristome and lid are almost pure white.

The upper pitchers of *Nepenthes eymae.*

Nepenthes inermis

From Sumatra, this is one of the strangest of pitcher plants, yet easy to grow. The unusual upper traps are only 1 or 2 inches (2.5–5 cm) tall, papery thin, funnel shaped with no peristome, and a thin, filamentous lid. They are pure green. It is believed that nectar on the tiny lid paralyzes small insects, which drop to the inner wall of the funnel trap and slowly slide into the digestive juices by means of a sticky, lubricating fluid.

Nepenthes inermis. The earliest evolved *Nepenthes* may have had similar sticky-cupped traps.

Nepenthes dubia

Similar to *N. inermis* and also from Sumatra; the primary difference is that the upper pitchers have a flat, ribbed peristome and can be orange.

Nepenthes glabrata

One of the daintiest and prettiest of the *Nepenthes*, a prize for any terrarium, this species comes from Sulawesi. The leaves are very narrow and lance shaped. The small lower traps are smooth and tubby, barely 1 or 2 inches (2.5–5 cm) tall. The peristome is yellow, with a small oval mouth and lid. The pitcher background is lemon green, delicately marked with some red streaks as though hand-painted.

The upper pitchers are similar but larger and more cylindrical. Sometimes tendrils and pitchers appear without leaves from the basal stems.

The dainty upper pitchers of *Nepenthes glabrata* seem almost hand-painted.

Nepenthes hamata

When people ask me which is the scariest-looking and most dangerous *Nepenthes*, I usually point to this one, which sends shivers down most animals' spines. The lower pitchers are long and narrow, heavily blotched in purple, and have prominent wings. The upper lid is hairy. It is the highly evolved peristome that is so disturbing, for the lip has transformed into a row of long, curved hooks, sharp as knives, that overhang the pitcher's mouth. The upper pitchers can be similarly colored to the lower traps or pure green, with particularly long peristome hooks—somewhat like a torture device from the Inquisition. One can only guess what this plant may be evolving into. Pray that it doesn't start walking.

The upper pitchers of *Nepenthes hamata* can send chills down animal's spines.

Mt. Kinabalu is Borneo's most famous and tallest mountain, over 13,000 feet (3,962 m) high. Many of the highland species I have discussed grow there, but there are a few *Nepenthes*, which follow, that grow nowhere else except on Mt. Kinabalu and its nearby peaks, or were first discovered on this mountain. These include some of the most notorious and beautiful of the tropical pitcher plants.

Nepenthes burbidgea

A lovely species, it is native to Mt. Kinabalu and the adjoining Mt. Tambuyukon. The lower traps are up to 8 inches (20 cm) long, ovoid and stout, with moderate wings and a broad peristome; the uppers are short, plump, and funnel shaped. The coloration of the uppers is spectacular, the background a pale yellow white, marked with sparse, irregular rosy blotches. The peristome is striped with red and pale yellow. The large lids are heavily spotted in purple. Burbidge, its discoverer, described them as "pure white, semi-translucent like eggshell, porcelain-white with crimson or blood-tinted blotches." I have found this species easy to grow with nightly lows around 60°F (16°C). At cooler temperatures, the plant is reluctant to pitcher.

Nepenthes villosa

From the higher elevations of Mt. Kinabalu, where nighttime temperatures can drop to 40°F (4°C), comes this popular, ground-scrambling species. Upper and lower pitchers are similar: plump and roundish, up to 8 inches (20 cm) tall, red-orange, and covered with an animallike pelt of fur. The lid is large and held horizontally. The spectacular peristome looks like a row of raised claws, sharp as razors and yellow to crimson. A seedling given to me grew for several years until a heat wave killed it. The best specimens I know of have been grown in refrigerated tanks! Clones from Mt. Tambuyukon grow at lower elevations than Mt. Kinabalu and may be easier to cultivate.

Nepenthes edwardsiana.

Nepenthes edwardsiana

From both Mt. Kinabalu and Mt. Tambuyukon, this species is very similar to *N. villosa,* but the pitchers lack the furry pelt and are long and cylindrical. The pitchers are golden to flushed red, the teeth of the peristome a similar series of raised flanges with downward-curved barbs. The long neck raises the lid far above the mouth.

Nepenthes macrophylla

Closely related to *N. edwardsiana*, this recently described species comes from Mt. Trus Madi, a neighbor of Mt. Kinabalu. The leaves are huge, up to 2 feet (0.6 m). The pitchers are more stout than those of *N. edwardsiana*, with a wide, gaping mouth. The teeth of the peristome are shorter and blood red.

The oozy "eggs" amid the lid bristles of *Nepenthes lowii.*

Nepenthes lowii

Discovered on Mt. Kinabalu by Hugh Low, this famous species also grows on several other tall peaks across Borneo. It may be the strangest of all *Nepenthes*, thanks to its bizarre upper pitchers. The lower traps are fairly normal and cylindrical, reddish brown, with a medium wide peristome. The oval lid is held horizontally and hints at peculiarities to come: under the lid hang many long, pointed appendages, like thin vegetable stalactites.

Caught in the act! Ch'ien Lee took this incredible shot of a tree shrew on *Nepenthes lowii*, a true crapivorous plant!

The upper pitchers, up to nearly a foot (0.3 m) long, look like weird, constricted gourds. The peristome is entirely lacking, the mouth wide and gaping. The pitcher suddenly narrows to an extreme waist, then balloons to a bulbous bottom. The exterior of the trap is mostly green, while the interior of the yawning mouth is shiny red to purple. The large lid of the upper pitcher is held vertically. It has the same strange, bristly projections as on the lower trap lids. In cultivation, *N. lowii* can be tediously slow to grow.

A strange mystery once surrounded this plant. Often, in the bristles of the lid, an oozy white substance is secreted, often taking the form of egglike beads. Professor J. Harrison, in the early 1960s, assumed they were snail eggs, and he reportedly saw small tree shrews eating them. It wasn't until plants entered cultivation that the "eggs" were discovered, by grower Cliff Dodd and myself, to be a product of the plant. Botanist Charles Clarke observed birds and shrews feeding on the "eggs," while their excrement fell into the pitcher! Since Clarke's revelation in the 1990s, this has been observed many times. *N. lowii* is a "crapivore"!

Nepenthes ephippiata

New to cultivation, this rare species from Borneo is most unusual, slightly resembling *N. lowii* and *N. rajah* (described next). The tubby pitchers are squat, slightly constricted at the waist, and about 6 inches (15 cm) tall. The large mouth has a narrow peristome. The main feature is the huge vaulted lid, the underside covered with peculiar short tendrils. The coloration is fabulous: the outer pitcher is pale crimson, the interior blood red. The lids are green with a red margin, turning fully red with age.

Nepenthes rajah

When Hooker described this species, also discovered by Hugh Low on Mt. Kinabalu, he wrote, "This wonderful plant is certainly one of the most striking vegetable productions hitherto discovered . . ."—and it remains so to this day. Also found on Mt. Tambuyukon, *N. rajah* grows along the ground as a scrambler. The large leaves are blunt and peltate, and the tendril originates from the underside of the leaf tip. The enormous pitchers are oval shaped, almost woody in texture, red to purple, with a large, gaping, oblique mouth.

Nepenthes rajah, still the largest tropical pitcher plant yet known.

The thick, fluted peristome is blood red. The interior of the tublike traps is pale green to pink and has no waxy zone, being entirely covered with large digestive glands. The giant lid is vaulted, red above and lime green below. The pitchers can be over a foot (0.3 m) in length, and can hold over a half gallon (1.9 L); specimens have been known to hold a full gallon (3.8 L). The flower spikes can also be impressive, standing as tall as 4 feet (1.2 m). Climbing stems are rare.

N. rajah was the first pitcher plant documented as having caught rats in the wild. Ch'ien Lee has recently photographed small mammals apparently licking nectar on the vaulted lids of this species, and their droppings may also benefit the plant.

Recent Discoveries and Introductions

As mentioned earlier in this chapter, the explosion of interest in growing *Nepenthes* in the past couple of decades has led to much exploration and discoveries of new species as well as old ones not seen for many decades. Most of these plants are found in the cooler tropical mountains of Southeast Asia where humans rarely—if ever—travel. In the days of old, humans would remove plants from the wild and attempt the slow propagation of them in horticulture, often decimating already rare populations. While poaching of plants is still a severe threat and must be condemned, the advent of modern tissue culture has allowed responsible horticulturalists to receive permits and obtain seed only, which then can be grown and multiplied in flask, enabling new discoveries to be available to the public within a few years. Often the most desirable can be rather expensive when first released on the market, but soon prices drop.

While many growers often desire plants with the largest pitchers possible, I equally enjoy the smallest, which are perfect for terrariums.

Nepenthes talangensis

First described by Joachim Nerz and Andreas Wistuba in 1994, from Mt. Talang in West Sumatra, this miniature can eventually have long climbing vines, which can be pruned back. Our plant at California Carnivores flowers from stems only 6 inches (15 cm) high. The tubby pitchers are heavily mottled in red on a pale olive background, with a large funnel-shaped bright red peristome. The pitchers are rarely much taller than

2 inches (5 cm). Upper traps are rather similar.

Nepenthes tenuis
One of the smallest species, with pitchers barely 1 1/2 inches (3.8 cm), it looks rather similar to *N. talangensis* but with a puffy-looking peristome that is nearly flat. The upper pitchers are funnel shaped and slightly larger, often rather purplish red.

Nepenthes pitopangii
A miniature with pitchers barely 2 inches (5 cm) in height. A single specimen was discovered in 2006 in Sulawesi. A second small population was found in 2010, nearly 60 miles (100 km) away.

Upper pitchers of *Nepenthes pitopangii*.

Nepenthes argentii
Discovered by British botanist George Argent in 1989 on Sibuyan Island in the Philippines, this tiny species grows on windswept, nearly inaccessible ledges near the summit of Mt. Guiting-Guiting. The chilly wind and low-growing surrounding vegetation usually prevent the growth of scrambling vines. The tiny pitchers are usually less than 1 inch (2.5 cm) high, tubby and dark olive with many purplish red blotches. The narrow peristome can be purplish black with tiny sharp teeth.

Nepenthes jacquelineae
Australian botanist Charles Clarke first described this West Sumatran species in 2001 and named it for his wife. The lower pitchers are up to around 2 inches (5 cm) tall; the uppers can reach 6 inches (15 cm) and are often deep red inside and out and funnel shaped. Its hallmark is the incredibly wide and flat peristome with hardly any ridges, glossy and glistening.

Nepenthes jacquelineae.

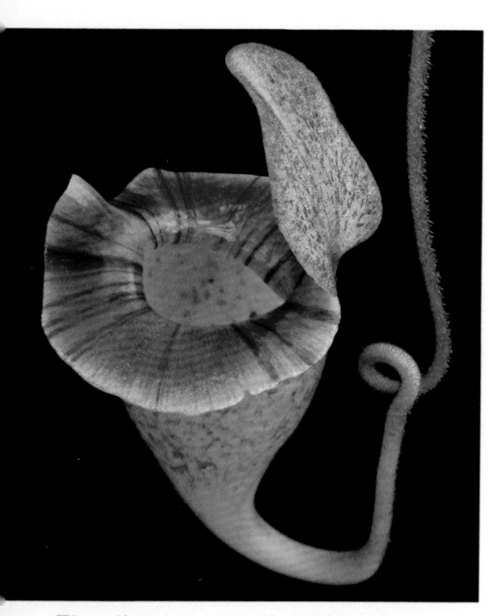

The slippery-smooth peristome dominates *Nepenthes platychila*.

Nepenthes platychila

Naturalist Ch'ien Lee discovered this species in the Hose Mountains of Sarawak in Borneo and described it in 2002. The pitchers of the fast-growing climbing vine are up to 6 inches (15 cm) long but usually smaller, very funnel shaped, yellowish green to pink, and mottled heavily in red splotches. The flaring peristome is smooth, flat, and usually candy-cane striped in green and red but can be nearly entirely red. Our specimen at California Carnivores pitchers readily, but the upper traps are rather short-lived.

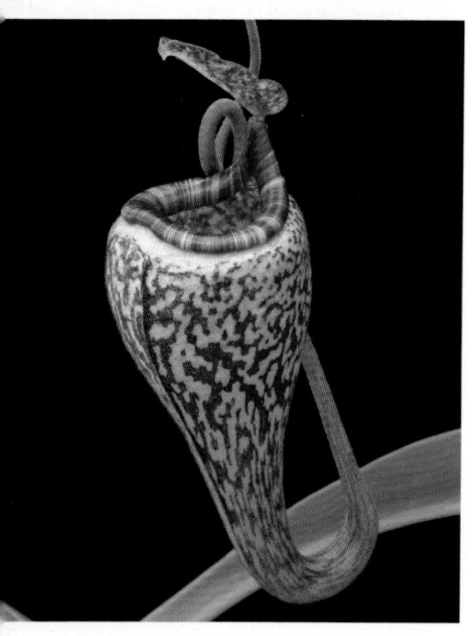

Nepenthes vogelii, newly described from Borneo in 2002.

Nepenthes vogelii

Another new discovery from the Hose Mountains. The lower pitchers are long and narrow like *N. fusca*, but its beautifully colored uppers are most striking.

The following two *Nepenthes*, one older and one newer, are most unusual in that the pitchers are curiously humpbacked, with the mouth actually on the side of the trap rather than on the top!

Nepenthes aristolochiodes

First described in Matthew Jebb and Martin Cheek's 1997 revised monograph, this West Sumatra species is critically endangered, thanks to poachers. The upper pitchers are up to 5 inches (12.5 cm) tall but usually much smaller. The pitchers are plump and olive green with much interconnected purple splotching, and the top of the trap is quite extraordinarily humpbacked, with a circular mouth along the upper front. The purple peristome dips into the pitcher, and the lid is also quite purple.

The bizarre humpbacked pitcher of *Nepenthes aristolochioides*.

Nepenthes klossi

This long-lost species was discovered in 1913 but not seen again until 2008, and it entered cultivation in 2012 thanks to tissue culture. Found in Papua, western New Guinea, the pitchers of the climbing vine can be nearly a foot (0.3 m) tall and are similar to species like *N. maxima*, being pale green with much purple spotting and a medium-size red peristome; however, the oblique mouth is on the upper front of the trap just below the humped top of the pitcher. The lid is large and shades the mouth, similar to *N. aristolochiodes*.

The following two species, recently discovered and published by Ch'ien Lee and associates, are quite bizarre.

The awesome upper pitchers of *Nepenthes jamban* look like toilet bowls!

Nepenthes jamban

This species' name comes from the Indonesian word for toilet, and it was described in 2006 after its discovery in North Sumatra. The spectacular pitchers can be up to 5 inches (12.5 cm) tall in the uppers, with a perfectly circular mouth and a lightly ridged flat peristome, often bright red. The gaping pitcher abruptly narrows to the tendril, not unlike a martini glass or toilet bowl.

Nepenthes lingulata
This odd species also comes from North Sumatra and has cylindrical, elongated lower pitchers to nearly 12 inches (30.5 cm) tall, the uppers half that height. They are richly colored a purplish black with a dark burgundy, flared peristome. But its most unusual feature is a thin, wiry appendage hanging over the mouth from under the lid, not unlike the lure of an anglerfish. This filiform projectile is baited with nectar glands. (*Lingula* means "small tongue" in Latin.)

Remote areas of the Philippine Islands have been a source of some fabulous new *Nepenthes* such as the ones that follow; more discoveries are no doubt to come.

Nepenthes sibuyanensis
Described by Joachim Nerz and associates in 1997, from Sibuyan Island's windy mountain slopes, this species is like a *N. ventricosa* on steroids with lower pitchers to 8 inches (20 cm) long. It has been used to produce some vigorous hybrids in cultivation.

Nepenthes copelandii
Discovered and described in the early years of the twentieth century, this species from Mindanao was rediscovered by Robert Cantley on the volcanic Mt. Apo in 1996. Vigorous and fast growing, this handsome plant has long, narrow pitchers that can reach 10 inches (25 cm) long, heavily streaked in purple on an olive background; its large teardrop-shaped mouth has a thin purple peristome. One I grew in a sunny windowed porch sent up a 7-foot (2.2 m) climbing vine in just a few months, the tendrils and pitchers festooned among strings of holiday lights.

Nepenthes attenboroughii and *N. palawanensis*
Among the many recently described tropical pitcher plants from the Philippines, these two received much attention in the general media when they were discovered. Both are among the largest ever found, the enormous pitchers rivaling those of *N. rajah* from nearby Borneo.

Named in honor of British naturalist Sir David Attenborough, *N. attenboroughii* was discovered by Stewart McPherson, Alastair Robinson, and Volker Heinrich on Mt. Victoria on the island of Palawan in 2007. The giant green pitchers are 12 inches (30.5 cm) tall and 6 inches

(15 cm) wide, with a circular lip heavily striped in green and purple, a large lid, and the interior of the voluminous trap mottled richly in purple.

In 2010, McPherson returned to the area to explore another nearby mountain, Sultan Peak, with several helpful guides and porters, and discovered *N. palawanensis*, very similar to the preceding species but with tubbier pitchers richly colored orange-red and even larger, with wavy, bristly wings. No upper pitchers were found, as the surrounding low vegetation was unable to support climbing vines.

The discovery of *Nepenthes attenboroughii* made headlines around the world.

Nepenthes robcantleyi

Possibly one of the most outrageously flamboyant *Nepenthes* yet discovered, two specimens were found by Robert Cantley in 1997 on Mindanao. Cantley—owner of Borneo Exotics in Sri Lanka, a tissue-culture lab and nursery specializing in tropical pitcher plants—had obtained permits to collect seed of *N. truncata*, a mostly lowland species also found in highland areas. Cantley germinated seed of this "black truncata" and chose three cultivars known as 'King of Spades', 'King of Hearts', and 'Queen of Hearts'. He grew these to maturity and successfully crossed them to produce more seed. Botanist Martin Cheek realized this was a new species closely related to *N. truncata* and described it in 2011.

The unbelievable *Nepenthes robcantleyi*.

The pitchers can reach over a foot (0.3 m) high with a fat, cylindrical body reddish brown to nearly blackish. Its most spectacular feature is the widely flared, fluted, deeply ridged peristome that can be 3 1/2 inches (8.5 cm), as deeply colored as the pitcher's body. Upper pitchers are not formed: this heavy, coarse plant can produce a thick vine not much more than 3 feet (0.9 m) long. A jaw-dropping spectacle in Cantley's award-winning displays at flower shows around the world, this species is currently becoming widely distributed, with plans for reintroduction into the wild, as the two original specimens disappeared after the location was logged.

Hybrids

Nepenthes, like *Sarracenia*, frequently hybridize in the wild, and artificial hybrids have been produced in horticulture for more than 150 years. Since the 1990s, with the resurgence of interest in growing the plants, the explosion of hybridization has been astronomical.

Unfortunately, CP lovers face problems with the nomenclature of *Nepenthes* hybrids similar to those of *Sarracenia.* The International Society for Horticultural Science (ISHS) has refused to sanction the use of grex names (plural, *greges*) for all the offspring of any artificially made cross (see page 354). They have only approved the naming of individual clones as cultivars, which can be registered through the International Carnivorous Plant Society (ICPS).

However, in 2008, several *Nepenthes* growers launched the International *Nepenthes* Grex Registry (INGR), where breeders can register greges. While not yet officially approved by the ISHS, it is a good beginning. You can go to the registry's website through links at www.alohanepenthes.com.

In the first edition of *The Savage Garden* I divided the sections on *Nepenthes* hybrids into "lowland" and "highland" plants, even though some hail from a combination of both habitats. Here I will discuss all hybrids together, beginning with the hybrids from the Victorian era and progressing through time to the modern era of new nurseries breeding hybrids. Where necessary I indicate whether a plant is natural, artificial, highland, lowland, or an intermediate combination, as well whether it is a vegetatively produced clonal cultivar or a grex or seed-grown cross.

Be mindful that not all seed-grown hybrids have "group" or grex names registered with the INGR.

Old Favorites

Nepenthes × *hookeriana*

This is the natural hybrid of *N. rafflesiana* × *N. ampullaria*; it has also been artificially produced. A large, vigorous plant, its lower pitchers have a heavy, squat, "boxy" look and are usually green with much red spotting. My favorites are those crossed with the red forms of *N. ampullaria*, which can be quite stunning.

N. × *hookeriana* was often crossed with the lowland *N. mirabilis* in the Victorian era, and many cultivars were named. Most have been lost to cultivation, but two fine ones still grown are *N.* × *coccinea*, with vibrantly red pitchers lightly flecked with green, and *N.* × *morganiana*, with fat, tubby green pitchers heavily mottled in red.

Nepenthes × *mixta*

A 1898 cross between *N. northiana* and *N. maxima*. The pale green traps are up to a foot (0.3 m) long and heavily streaked in red, and the large, slanted mouth has a wide, luscious peristome that glistens bright red. Upper pitchers are equally impressive but more funnel shaped, like giant cornucopias. Two more richly colored clones are *N.* × *mixta sanguinea* and *N.* × *mixta superba*. This is an exceptional Victorian hybrid produced by Mr. Tivey at Veitch Nurseries and is still extremely popular. It does best in lowland conditions.

Nepenthes × *dyeriana*

Released by Tivey in 1903, this is a cross of *N.* (× *mixta*) × (*rafflesiana* × *veitchii*). The heavy pitchers reach 14 inches (35.5 cm) and are green with many red/purple/brown streaks. The large peristome is candy-cane striped, turning bronze in good light. Like × *mixta*, this was produced at Veitch Nurseries and does best in similar growing conditions.

Nepenthes × *trichocarpa*

A natural cross between *N. gracilis* and *ampullaria* and was once thought to be a species. It is excellent in warm terrariums, producing clusters of tubby, spotted lower pitchers resembling fat *N. gracilis* traps. I'd love

Victorian hybrids, left to right: *Nepenthes* x *dyeriana*, *N.* x *mastersiana*, *N.* x *morganiana*, *N.* x *coccinea*, *N.* x *hookeriana*.

Nepenthes x *tiveyi*.

to see an artificially produced remake using dark-colored *N. gracilis* and red forms of *N. ampullaria*!

Nepenthes × _mastersiana_
Produced by Court for Veitch Nurseries in 1883, this was one of Sir Harry's favorites (and mine!). A highland cross between *N. khasiana* and *N. sanguinea*, it has large cylindrical pitchers with an oval mouth and lid, and the coloration is vibrant, especially in the clone, *N.* × *mastersiana purpurea*.

Nepenthes × _tiveyi_
An utterly beautiful Veitch Nurseries hybrid between *N. maxima* × *veitchii*. It has large, colorful pitchers with a flared rainbow peristome that turns orange to red. An intermediate plant that can take cooler or warmer conditions.

***Nepenthes* 'Lieutenant R. B. Pring', *N.* 'St. Louis', and *N.* 'Henry Shaw'**
From 1918 to 1956, the Missouri Botanical Garden housed one of the largest *Nepenthes* collections in the world, under the direction of George H. Pring. Three cultivars were named from a cross of *N.* [(*rafflesiana* × *hirsuta*) × (*rafflesiana* × *ampullaria*)] × (*rafflesiana* × *hirsuta*). All are lowland and have tubby, squat pitchers of similar shape, but with coloration differences.

***Nepenthes* 'Ile de France'**
A hybrid from France is noteworthy. A natural hybrid dubiously believed to be *N. mirabilis* × *thorelli* (*N. thorelli*, not seen for nearly a century but

rediscovered in Vietnam by Francois Mey in 2011) was introduced by famous nurseryman Marcel Lecoufle and named *N.* 'Lecouflei'. He crossed this plant with *N. mixta* var. *sanguinea* and introduced the beautiful *N.* 'Ile de France'. The cylindrical lower pitchers are flushed pink in the upper part, with streaks of chocolate-red, and the flat peristome is striped green and crimson; the upper pitchers are paler. This cultivar does best as a lowlander.

Nepenthes × *rokko*

Japanese growers produced many hybrids in the early decades of the twentieth century, but the hard winter of 1940 and the war that followed left few surviving plants. Hybridization of *Nepenthes* resumed around 1950 after the establishment of their society, and growers give their crosses grex or group names, so the offspring may be variable. I have grown and distributed some of these in the West.

One such hybrid is *N.* × *rokko*, a cross between the alleged *N. thorelli* × *maxima*. Lower pitchers reach 8 inches (20 cm) with a scalloped red peristome and many red streaks and spotting. Upper pitchers are funneled, green, with light spotting. It does well in both highland and lowland conditions, and in time can produce massive amounts of climbing vines that can grow to 7 feet (2.1 m) long in a few months.

Koto hybrids

In the 1970s and 1980s, Dr. K. Kawase of the Kosobe Conservatory at Kyoto University created dozens of hybrids and was the first to use many highlanders, which required less heating expenses. He produced 163 hybrids between 1973 and 1983, and named them alphabetically after Koto, which means "old capitol," referring to the city of Kyoto. Thus we have greges named × *aglow koto*, × *balmy koto*, × *dreamy koto*—all the way to × *zonal koto*; some of these are grown in the West.

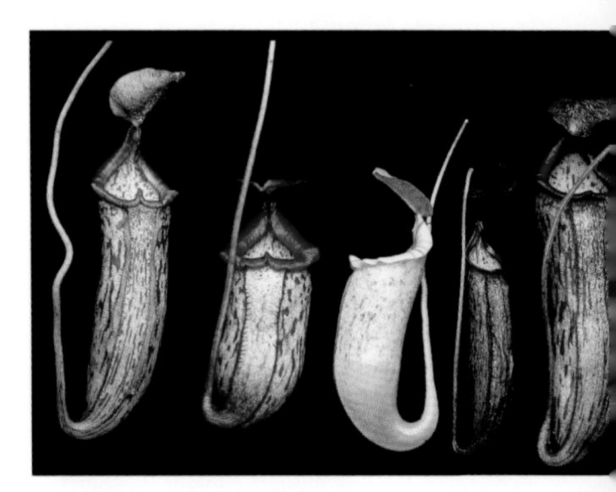

Related to highland hybrids, left to right: *Nepenthes* x *rokko*; lower and upper pitchers of *N.* (x *rokko*) x *thorelli*; *N.* 'Santa Mira' var. 'Jack Finney'; and *N. thorelli* x (x *wittei*).

The New Wave

"New Wave" is usually used to describe a movement of originality and change in film or music, such as the French New Wave cinema of the 1950s, or the New Wave in rock music that occurred in the late 1970s and early 1980s. I can think of no better description to explain what has happened in growing *Nepenthes*—particularly hybridization—in the early twenty-first century. And much of this has to do with just a handful of retail and wholesale nurseries who have the facilities to grow plants to maturity, cross them, and reproduce them *en masse* via tissue culture, seed, and cuttings. The new wave of *Nepenthes* breeders in the twenty-first century has been producing a stunning variety of modern hybrids using not only familiar species but also the incredible new ones recently discovered.

Be mindful that the rarest of *Nepenthes* hybrids are the clonal cultivars I have just reviewed previously in Old Favorites. These plants can be propagated only by taking cuttings of the vines, since the plant tissue taken from the cultivars cannot be successfully sterilized without killing the plant cells.

On the left, *N.* 'Miranda'; on the right, *N.* (*ventricosa* x *alata*) x *alata*, often sold as *N. alata*.

However, there are currently only two large *Nepenthes* that have been mass-produced this old-fashioned way of rooting cut vines, *N.* 'Miranda' and *N.* (*ventricosa* × *alata*) × *alata*. They are grown and propagated in massive greenhouses and wholesaled to the retail market from Paris to Miami to Hong Kong. Both of these hybrids are usually sold in large pots of dense sphagnum peat or heavily washed coco peat, and they do best when eventually moved to larger pots in a more typical *Nepenthes* soil.

Nepenthes 'Miranda'

This plant is a cross between *N.* (× *mixta*) × *maxima*. The huge lower pitchers can reach 8 inches (20 cm) or more;

they have large bristly wings, much red streaking, and a wide, fluted red peristome—a classic-looking tropical pitcher plant. The upper pitchers are green cornucopias but retain the blood-red lip. It can be grown outdoors in tropical to subtropical conditions, but stops or slows growth during winter chills. Indoors, choose a humid and partly sunny area—this one gets huge!

Exotica Plants nursery specializes in *Nepenthes* hybrids, such as *Nepenthes lowii* x *truncata*.

Nepenthes (ventricosa × alata) × alata
This plant is marketed under the name *N. alata* and doesn't have a clonal name. Much less gaudy and colorful than *N.* 'Miranda', it is a vigorous grower and can do quite well in humid indoor areas of very bright to partly sunny conditions. Both upper and lower pitchers are 3 to 4 inches (7.5–10 cm) long, smooth and waisted with a bulbous bottom, and green to pinkish red when grown in higher light levels.

Robert Cantley's Borneo Exotics, a wholesale export nursery located in Sri Lanka since 1997, has been one of the most productive and successful outlets. Their website, www.borneoexotics.com, offers links to their retail vendors around the world. They propagate plants by both tissue culture and seed. Winners of numerous awards for their displays, from the Chelsea Flower Show in London to the Singapore Garden Festival, they have introduced more newly discovered species than any other mass-producer in the world, and the number of hybrid selections continues to grow. One of the most

Exotica's *Nepenthes* (*ventricosa* x *sibuyanensis*) x (x *trusmadiensis*).

popular, *N.* 'Lady Pauline', a cross of *N. talangensis* × *maxima*, is a beautiful plant with a large purple peristome (see photo on page 293). They have also been producing smaller hybrids suitable for terrariums such as *N. ventricosa* × *aristolochiodes*. While not a hybrid between species, they crossed 'Queen of Hearts' × 'King of Spades', creating a stunning version of *N. robcantleyi*.

Exotica's *Nepenthes (ventricosa* x *sibuyanensis)* x [*(spectabilis* x *talangensis)* x *truncata*].

Exotica Plants (www.exoticaplants.com.au) in Queensland, Australia, owned by Geoff and Andrea Mansell, has been specializing in creating fantastic hybrids of tropical pitcher plants for nearly two decades. They sell only seed-grown plants directly to the public but also have a few mass-produced for wholesale. Exotica Plants has incredible breeding stock they use to produce popular hybrids, such as giant forms of *N. merrilliana* and *N. truncata*, as well as miniature species like *N. campanulata*.

Hawaiian nurseryman Sam Estes has produced many hybrids, such as *Nepenthes* 'Song of Melancholy'.

Sam Estes, in Pahoa, Hawaii, also grows and wholesales seed-grown hybrids, most given grex names. His website, www.alohanepenthes.com, has links to retail outlets distributing his creations. The nursery Leilani Hapu'u has produced many beautiful hybrids, often crossing lowland and highland species to produce intermediate plants that do well at "room temperature," requiring a minimum of heating and cooling. Since these are seed-grown greges, they can be variable. Another bonus with Sam's plants is that they are often a few years old and large. Sam manages the International *Nepenthes* Grex Registry.

Two other nurseries also produce both rare species and intriguing hybrids. Malaysiana Tropicals, a Borneo nursery producing mostly tissue-cultured plants, can be found at www.malesiana.com. A European tissue-culture lab and nursery owned by Kamil Pasek in the Czech Republic can be reached at www.bestcarnivorousplants.com.

Nepenthes 'Peter D'Amato', a hybrid between *N. ventricosa* x *lowii*, named after the author by Bill Baumgartl.

Nepenthes 'Peter D'Amato'

It's always nice to have a beautiful plant named after one's self. Johannes Marabini of Germany crossed *N. ventricosa* × *lowii* and called the grex *N. briggsiana*, distributing the seeds to growers around the world. Dr. Bill Baumgartl chose one clone and named it after me. Gourd-shaped, deep red, with a finely toothed peristome, Barry Rice has described it as "grotesque." I am honored!

CULTIVATION

(See Parts One and Two for further details.)

SOIL RECIPE: *Nepenthes* enjoy loose, open soil that remains wet to moist but allows drainage of excess water. They are tolerant of a wide variety of soil mixes. The best include a portion of long-fibered sphagnum, the rest of the medium being a combination of coarse materials. My deluxe recipe is about 70 to 80 percent New Zealand or other high-quality long-fibered sphagnum moss; the rest of the mix is a combination of fine orchid bark, perlite, and tree fern (osmunda) fiber. Another good mix is one part long-fibered sphagnum to one part tree fern fiber or orchid bark. Some growers have had success with pure shredded fir bark, and so have I, but I prefer the bulk of the mix to be a high-quality sphagnum moss.

continued

Cultivation, *continued*

CONTAINERS: All *Nepenthes* containers must have drainage holes. Place a thin layer of sphagnum at the bottom of the pot to prevent the gradual loss of soil through the holes. Plastic pots work well, as do glazed ceramics. Even better are wooden boxes or orchid baskets. Avoid metal zinc baskets, which poison *Nepenthes*. Four-inch (10 cm) pots suit young plants; 6- to 10-inch (15–25 cm) pots (or larger) suit mature plants.

WATERING: In greenhouses, avoid the tray system entirely and place the containers on benches or hang them so that water can freely drain away. In terrariums and on windowsills, place the pot in a shallow saucer and water from overhead as soon as the water in the saucer evaporates. Don't allow the pot to sit in deep water for extended periods. Greenhouse plants should be watered every day or before the soil dries out. Always water from overhead. If the medium dries out, the pitchers may shrivel and brown very quickly, even if the leaves and stems survive.

LIGHT: Most *Nepenthes* enjoy very bright, diffused light or partly sunny conditions. Lowlanders often can take bright shade. Greenhouses generally require 50 percent shading.

CLIMATE: All *Nepenthes* are tropical plants, roughly divided into lowlanders and highlanders. Lowlanders require temperatures in the 80s°F and 90s°F (27–35°C) during the day, 60s°F and 70s°F (16–24°C) at night. Colder temperatures, even briefly, may stunt or kill them. Highland species require temperature drops at night. Highlanders do best in the 70s°F and low 80s°F (24–27°C) during the day, 50s°F and low 60s°F (10–16°C) at night. Many highlanders tolerate brief nighttime drops to the 40s°F (4°C), as long as daytime temperatures rise. Exceptions are mentioned under the species listing. Humidity must be high—above 60 percent—all of the time. Highland plants can handle more humidity fluctuations, with the highest humidity at nighttime.

DORMANCY: *Nepenthes* require no dormancy but may slow down in winter months.

OUTDOORS: If you live in a tropical climate similar to their native habitats, *Nepenthes* make wonderful potted outdoor plants, especially near latticework or trellises in partly sunny areas where they can climb. In humid subtropical or warm temperate climates, outdoor growing can be tricky, due to seasonal fluctuations, so plants are best moved indoors or to greenhouses for winter. In places like southern Florida, many lowlanders and hybrids succeed year-round outside, but you must protect them during rare winter chills. Along the coast of California, in the frost-free fog belt, many highland species can thrive. As a rule, they despise frosts and periods of hot temperatures with low humidity.

BOG GARDENS: *Nepenthes* are not suitable in bog gardens.

WINDOWSILLS: A surprising development in recent years has been experimental growing of *Nepenthes* on windowsills—with often wonderful results. Bright light to partly sunny conditions are necessary, and high humidity with frequent misting is helpful. Of the many plants found to thrive in good conditions, I can recommend *N. ventricosa* as the first to try. Some lowland hybrids with highland ancestry are also possible, such as *N.* x *dyeriana*. See the individual listings of species for further recommendations. If the plants don't pitcher, low humidity and light are usually why. The tendrils of upper pitchers will need something to grasp.

TERRARIUMS: Larger tanks are an excellent way to grow *Nepenthes*. Lowlanders may require heating pads to maintain a 60°F to 70°F (16–21°C) minimum. Highlanders will thrive in homes that are chilly at night: 50°F to 60°F (10–16°C). Refreezable ice packs placed in a tank overnight can cool highland plants considerably. Choose smaller species, and prune back extensive climbing stems. Lowlanders can take warm and steamy tanks; highlanders appreciate good air circulation and misting at nighttime.

GREENHOUSES: *Nepenthes* grow best in greenhouses. Lowlanders require stove houses or hothouses; highlanders do best in warm houses, but some tolerate cool houses. Many can be grown together at roughly 60°F (16°C) minimum, 85°F (29°C) maximum. While many highland varieties can tolerate hothouse conditions, lowlanders can be damaged at cooler temperatures. Hybrids are much more tolerant of temperature fluctuations, but this is heavily influenced by their parentage.

FEEDING: Any suitably sized insects can be fed to these pitcher plants, such as crickets, sow bugs, and mealworms. Dried insects are also excellent.

continued

Cultivation, *continued*

FERTILIZING: *Nepenthes* appreciate fertilization. During the warmer months, apply twice monthly. In winter, once a month will suffice. Apply as a foliar feed. Heavily dilute the fertilizer before applying. Avoid Miracid, which can stunt many *Nepenthes*.

TRANSPLANTING: Very old plants that need repotting should have all tall climbing stems removed. Discard old soil and trim away excess roots, then soak in SUPERthrive and repot. Older plants may take a while to recover; a foliar fertilizer can help.

GROOMING: I usually remove old pitchers when they are halfway brown and cut off old, dead leaves. Mature *Nepenthes* can survive many years in large pots but will require pruning of larger stems. This will also encourage new basal shoots to develop, and the cuttings can be used for propagation. Typically, mature plants develop one or more basal shoots annually. These form ground rosettes for one or two years before beginning to climb. It you prefer lower-growing plants with ground pitchers, pruning stems will not harm the plant. Never remove a climbing stem before a basal shoot has developed. Climbing stems require something to climb: other plants, hangers, pipes, lattice, and so on.

PESTS AND DISEASES: The primary pests of *Nepenthes* are thrips, scale, mealybugs, and aphids. Systemic and contact insecticides work best. Flea collars help in terrariums. Sometimes the plants are bothered by a fungus that causes rusty spots on the leaves. Treat with a fungicide.

PROPAGATION

Nepenthes are easy to propagate by seed and stem cuttings.

SEED: To produce seed, male and female plants need to be in flower at the same time, or pollen can be stored in foil packets in the freezer for up to one year for future use. When they bloom, individual flowers open several at a time, working their way up the spike over a few weeks' duration. To pollinate, transfer pollen from a male flower to the stigmas of females. One method is to remove a ripe male flower by clipping it and, using forceps, dab the pollen onto the stigmas. Alternatively, one can shake the male spike over a sheet of plastic or aluminum foil. Ripe pollen will fall and can be collected with a small paintbrush and transferred to the female. Repeat pollination to ensure good seed set. When you're successful, the female ovaries will swell over several weeks, turn brown, and crack open, revealing seed. Be sure to label flower spikes that have been pollinated, as well as any stored pollen.

Nepenthes seed is short-lived and should be sowed as soon as possible. It can be stored for several weeks in the refrigerator (do not freeze seed), but be aware that this can kill seed of lowland species. Good mediums to sow seed onto include milled sphagnum or a mix of peat and sand or perlite. Keep the soil damp and sow sparsely. High humidity, as in a covered seed tray, is required. Keep the seed in an environment similar to that of the parents, but out of the direct sun. Grow lights are useful.

After the seed has germinated, remove the cover. Allow seedlings to grow until small rosettes are formed (six months to one year). Gently remove them and transfer to small individual pots of their preferred soil recipe. Fertilizing should be done with care, using a heavily diluted solution the first year, once monthly.

CUTTINGS: Propagating *Nepenthes* from stem cuttings is the fastest way to obtain large plants, and the only way to multiply cultivars.

Along the climbing stem, and adjacent to each leaf base, are dormant nodules or shoots. Occasionally these grow on their own, producing branching stems. If the growing point of a stem is removed, nearby shoots start to grow, replacing the removed portion.

Some growers take cuttings a little at a time, removing the growing point for propagation and waiting until new shoots sprout before removing that next section. Alternatively, a whole climbing stem may be removed and divided into sections. Sections of the stem should have from one to three leaves attached. If multiple leaves are attached, remove the lowest. Remaining leaves should be cut in half. The lower portion of the cut stem is then treated with a fungicide/rooting hormone such as Rootone (a powder) or Dip'N Grow (a liquid), following the manufacture's instructions. Insert the cuttings into pots of medium. A good medium to use is pure long-fibered sphagnum. Rock wool is superb for rooting *Nepenthes*. Once the cutting has rooted, the entire block of rock wool can be planted in soil. Keep the medium damp at all times and place the potted cuttings into a propagation case,

continued

terrarium, or similar enclosed tank to ensure very high humidity. Place in bright light, but out of direct sun, at a temperature range similar to that of the mother plant. Mist frequently, keeping the medium damp.

Within weeks to months the cutting should root, and the dormant bud sprout. It can then be transplanted and grown normally. Not all cuttings may survive. Remove dead ones from your propagating case. SUPERthrive and fertilizers can be used once the cutting has rooted.

There is a hormone available to promote growth of dormant shoots on plants, and commercial products are often advertised in orchid magazines, such as *Orchid Digest*. One such brand is Keiki Grow. An Internet search will bring up vendors. These shoot promoters are usually pastes applied to the dormant bud. They are helpful to promote new shoots on a stem, which can later be removed as cuttings to be rooted. They can also be applied to cuttings with dormant buds, in addition to rooting hormones along the cut stem.

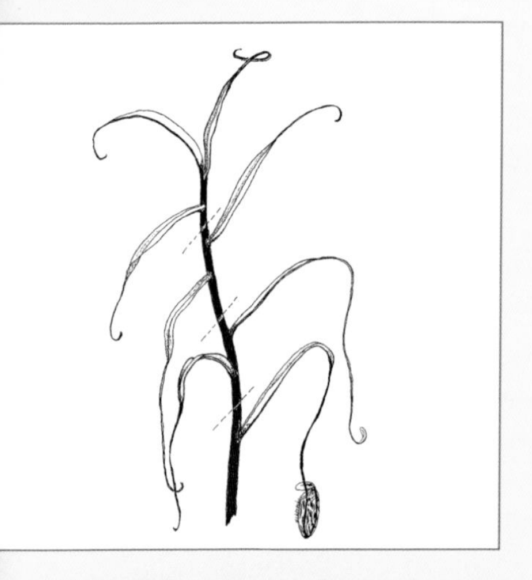

Cutting a *Nepenthes* vine for propagation.

Rooted cuttings with emerging shoots, ready to be potted.

Some vigorous *Nepenthes* can be rooted in plain pure water, if all other conditions (such as high humidity) are right. Change the water frequently.

There are other ways to propagate stem-growing plants like *Nepenthes*, such as layering and air layering. Consult good books on gardening and houseplants for more techniques such as these. The preceding methods are still the most popular.

TISSUE CULTURE: Tissue culture works well using seed as the generating source, and it has recently resulted in many once rare species becoming common and affordable. Vegetative tissue culture is still being perfected.

Other Savage Plants from the Demented Mind of Mother Nature

Roridula gorgonias.

Not all carnivorous plants are showy, ornamental specimens, and not all are as popular as the Venus flytrap or *Nepenthes*. A few, in fact, may not fit the stricter definitions that define a true carnivore—namely, they are plants that don't produce their own digestive enzymes. For example, *Heliamphora, Darlingtonia*, and *Byblis liniflora* may lure, catch, kill, and absorb insect prey, but they do so with the help of bacteria or other life forms to aid in digestion. Only very recently, with the help of better technology, have botanists begun to get a hint of enzyme production in carnivorous plants long suspected of having none. So what about other plants long dismissed as noncarnivorous even though they kill insects "defensively"—the petunias, potatoes, catch-fly, and other sticky plants? Are they carnivorous, semicarnivorous, or subcarnivorous? The answers may elude us, as the distinction may not be as black-and-white as botanists would hope. Some species may be entering or leaving a carnivorous phase in their long history of evolution. A sudden mutation of this or that gene, or an ecological cosmic catastrophe from outer space, may quickly speed up evolutionary jumps and turn a petunia that uses sticky hairs to defend itself against bugs into a petunia that eats those bugs.

In this chapter I will review some of the more obscure plants of the savage garden. A few of these may fit the full definition of true carnivorous plants; others may be subcarnivores that digest by proxy. Some may be less popular because they appear less dramatic or because you would need a microscope to see the drama unfold. Most are still sought after by the hard-core collector, but you'll probably never see them at your local garden center. The majority are easy to grow, but one or two can be rather challenging. While all are in cultivation, a few are so rare that live plants in a pot may be counted on one hand.

The Waterwheel Plant

(ALDROVANDA)

The waterwheel, *Aldrovanda vesiculosa*, is a true aquatic carnivore that is currently believed to be a direct relative of the Venus flytrap.

I have always been impressed by young children who visit my carnivorous plant nursery and eagerly ask where the *Aldrovanda* are. When

I show them, their happy little faces become shadows of disappointment. "I thought it was a lot bigger," they say, as I hand them a magnifying glass. I tell them that the plant they saw on TV or in a science book was no doubt photographed under a microscopic lens.

But *Aldrovanda* is a fascinating little plant. It was first discovered in India by Leonard Plukenet in 1699. In 1747, Italian botanist Gaetano Monti described it as *Aldrovandia* from plants found near Bologna by Dr. Carlo Amadei. Linnaeus misspelled it in 1753 as *Aldrovanda*, listing it in his *Species Plantarum*.

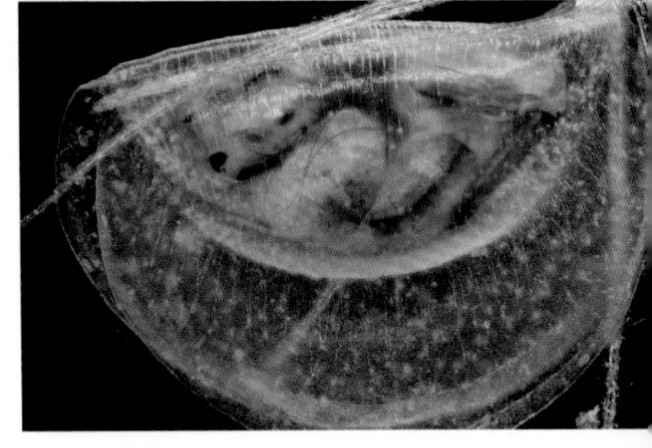

An *Aldrovanda* trap with prey.

Waterwheel plants are fond of placid, acidic ponds and lakes throughout much of the world. They are found in scattered and often isolated populations in Europe, Africa, India, Australia, and Japan (where it recently became extinct in the wild). In Europe, the plants are vanishing due to pollution. The plants float just below the surface of water, often amid reeds, cattails, and other water-loving plants. In temperate areas the plants go dormant as tightly rolled turion buds in winter, whereas in the tropics they grow year-round.

Waterwheels are rootless. They typically have stems about 4 to 6 inches (10–15 cm) long, one end growing while the other dies away. The leaves appear in whorls of six to nine. Each leaf has a broad petiole that ends in a small trap. Around the trap are long, pointy bristles, believed to prevent the trap from being damaged by other vegetation as the plant freely floats about in the water. The whorls of leaves are compactly grown along the stem, a dozen or more along its length. Small, single, white-petaled flowers appear above the water's surface in summertime.

The tiny trap is barely 1/12 inch (2 mm) across and remarkably like that of a Venus flytrap. It looks like a translucent green clamshell. Along its free margins, each lobe is lined with numerous tiny hooked teeth. Inside the trap are long, filamentous trigger hairs, about forty in each trap. The inside of the trap is liberally peppered with digestive glands.

Tiny swimming creatures, like water daphnia and eelworms, enter the trap for unknown reasons. No lure has been detected. Regardless, when one or more trigger hairs is touched, the trap quickly shuts. The tiny teeth interlock, imprisoning the prey. Slowly the margins of the trap squeeze together, until the prey is forced to the base of the trap near its hinge, and most water is expelled. The trap seals itself with a viscid secretion. Then digestive enzymes and acids are secreted, and the victim dissolves and is absorbed. Each trap can catch several meals.

Waterwheel plants, once rare in cultivation, have become more common in recent years as cultivation methods have improved. I have never succeeded with tropical forms from places like northern Australia, because when I attempt them in my hothouse, algae—the scourge of *Aldrovanda*—overwhelms and kills them. We have found temperate forms from Europe and Japan rather easy, however, when grown outdoors in a correct environment.

At California Carnivores we use shallow plastic containers similar to busboy tubs from restaurants. At the bottom of this tub we add a couple of inches (5 cm) of wet sphagnum peat moss to which we then add about 6 inches (15 cm) of pure water. It appears that *Aldrovanda* requires a high quantity of CO2 produced by other water plants they grow with, so we add rushes (*Juncus* spp.), sedges (*Carex* spp.), cattails (*Typha*) or free-floating water hyacinth. The first three plants can be in pots nestled into the peat at the bottom or with the roots inserted into the peat. We allow the peaty water to settle for a few days before adding the waterwheel plants. Never allow the water level to get much deeper than described here; in the wild the plants usually float along the shallow edges of ponds amid other vegetation.

Waterwheel plants floating in a tub.

To control algae, we have found that water fleas, mosquito larvae, and even tadpoles help, and they usually avoid eating the waterwheels, preferring instead to feed on algae growing on the water plants. Water fleas are also great food for *Aldrovanda*. A screen placed over the tub will trap adult mosquitoes and keep out pests like raccoons.

Our tubs sit in full sun, and in winter the waterwheels go dormant and are exposed to frosts. It is important to never let the water freeze solid. *Aldrovanda* can be propagated by simply breaking the stems apart.

Two Carnivorous Bromeliads

(*BROCCHINIA REDUCTA* AND *CATOPSIS BERTERONIANA*)

Bromeliads are a large order of plants found in tropical and subtropical regions of the Americas. A few varieties have become popular garden-center and grocery store plants. Bromeliads can grow epiphytically on trees or terrestrially on the ground. Their leaves are usually held in a tight, upward rosette, the base of the leaves forming a tank where rainwater is collected and held. It has always been assumed that bromeliads benefit from leaf debris and the occasional insect; that when these fall into the bromeliad's water-holding tank, they break down and decompose, supplying the plant with extra minerals. But it wasn't until recently that evidence was offered showing some of these plants had taken the next step or two forward along the path to carnivory.

In 1984, Professor Thomas Givnish showed that *Brocchinia reducta*, a terrestrial bromeliad from the Guyana Highlands of South America, showed features of a carnivorous plant. Not surprisingly, this species shares its home with *Heliamphora*, growing on and around the tepui table mountains of Venezuela and Guyana.

Brocchinia reducta give off a sweet, honeylike aroma that lures bugs. No nectar is actually produced. The leaves are coated with a wax that easily crumbles under insect feet, making keeping a foothold difficult. Insects fall into the pool of water the leaves contain; there they drown and are broken down by bacterial action. Hairlike trichomes at the base of the tank readily absorb the minerals of the dissolved prey.

Two carnivorous bromeliads. On the left, *Catopsis berteroniana*; on the right, *Brocchinia reducta*.

Since frogs enjoy making their homes in the plant, as is frequently seen in both cultivation and in the wild, I wouldn't be surprised if they help digest prey for the plant: perhaps the frogs' fertilizer-rich droppings are a reward to the plant for providing a good home, much the way I suggested for *Heliamphora* (pages 142 to 143).

Brocchinia reducta, growing in sunny wet grasslands or high on the tepuis, usually has tightly held leaves of a lovely golden yellow. In cultivation, the leaves are usually more openly held and more grayish green. The plants reach about 14 inches (35.5 cm) tall. The branching spikes of flowers are inconspicuous, but when the plants do flower, as with many bromeliads, the mother plant dies and is replaced by numerous offshoots or pups. *Brocchinia hechtioides* is nearly identical, and the two species are often found growing together.

A second bromeliad, *Catopsis berteroniana*, has also recently been described as insect-eating. It grows as an epiphyte on trees and is found in Florida and parts of Central and South America. This species produces a powdery wax so abundantly that its open crown of pointy leaves can appear almost white. It works in the same manner as the *Brocchinia*.

I have found both species to be very easy to grow. They succeed outdoors in warm temperate and subtropical climates that are frost free. *Brocchinia* does well in pots of soil that are two parts peat to one part each of sand and perlite (or lava rock). *Catopsis* grows well attached to branches or sitting in a pot of fine orchid bark and lava rock. Cool houses and warm greenhouses are perfect for them. Don't grow them on the tray system; instead, sprinkle them frequently from overhead with water, keeping the soil damp and well drained. Some heavily diluted fertilizer about once monthly will greatly benefit them. Propagation is best by division of their gradually produced clumps, but don't remove any offshoots until they are at least one-quarter the size of the parent. Tissue-cultured plants are becoming available, so these uncommon species should soon become more popular.

The Corkscrew Plant

(GENLISEA)

Corkscrew plants are extremely odd carnivores that entered cultivation in the 1980s. They were so rare that there was no common name for the plant until Adrian Slack used the term "Forked Trap" in his books at the time. In the first edition of *The Savage Garden*, I coined the common name "Corkscrew Plant," which is now in common usage. There are estimated to be around thirty species, some still to be officially described, growing in tropical Africa, Madagascar, and South America. They are closely related to *Pinguicula* and particularly to *Utricularia*, although the traps of *Genlisea* are unlike those of any other carnivorous plant in the world.

Genlisea grows in habitats similar to those favored by bladderworts and even sundews, and often grows alongside them. They grow as terrestrials or semiaquatics in wet to waterlogged peaty sand. All the species are very similar in leaf and trap structure, so identification usually relies on flower structure, which is rather similar to that of bladderworts.

Genlisea as grown by Geoff Wong.

The plants produce small rosettes of spade- to strap-shaped leaves flat on the soil surface, from 1 to 3 inches (2.5–7.5 cm) across. The strange traps are underground, about 2 to 6 inches (5–15 cm) long. They resemble a two-pronged fork or corkscrew. A cylindrical stalk extends downward from the base of the plant. Midway down, this swells into a hollow, bulblike digestive chamber, the "stomach" of the trap. It continues downward as a tubelike structure, then abruptly branches into two corkscrewlike appendages. At the base of the fork where it branches is a slitlike mouth, which spirals all the way down both prongs of the trap.

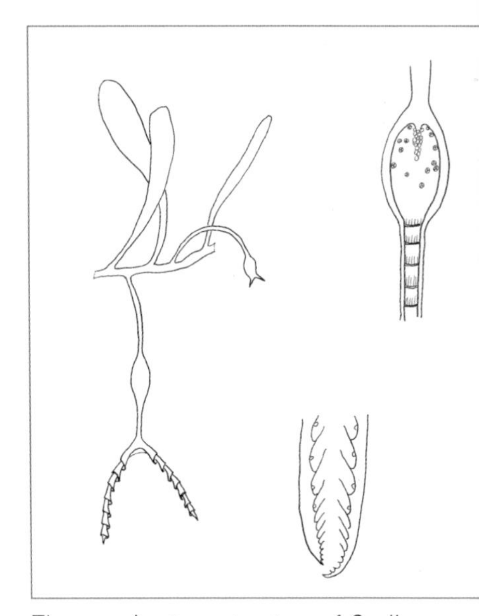
The amazing trap structure of *Genlisea*.

Tiny creatures, similar in size to the prey of waterwheel plants and bladderworts, can enter this slit at any point along its length. What lure, if any, attracts the prey is presently unknown. There is some evidence that a vacuumlike suction helps draw in victims, but other researchers disagree and think some chemical lure may attract prey. Once within the slit, these tiny creatures find themselves in a tubelike tunnel. They cannot escape because the slit is lined with bristly hairs that force them to continue upward along the tube of the prong. This one-way journey to death leads them all the way up the bristle-lined tube and into the digestive chamber. Imprisoned, the prey are dissolved by digestive juices.

Genlisea are simple plants to grow. Since they are tropical, minimum temperatures should be 60°F (16°C). They enjoy waterlogged conditions, so I prefer to grow them in undrained containers 6 inches (15 cm) deep, such as brandy snifters. The soil should be pure sphagnum peat or peat with some sand or long-fibered New Zealand sphagnum moss, filling the container nearly to the brim. Keep the soil wet to flooded. Grow them in very bright to partly sunny conditions. They do nicely in hothouses and warmer terrariums.

Unfortunately, while the plants are simple to grow in the above manner, you cannot see the traps! Geoff Wong won "Best of Show" at the San Francisco Flower Show for his *Genlisea*, shown in the photograph on page 341, which was a very clever way to cultivate the plant in a way that made the traps clearly visible. The rosette of the plant grows in sphagnum moss in a box that sits on a clear plastic container filled with water. The bottom of the box can be any plastic screened material that will hold the sphagnum while allowing the traps free access into the water below. Plastic strawberry baskets from grocery stores are a good source of screening to hold the moss. The traps will grow downward from the moss box into the transparent water container.

A corkscrew plant, revealing the underground traps.

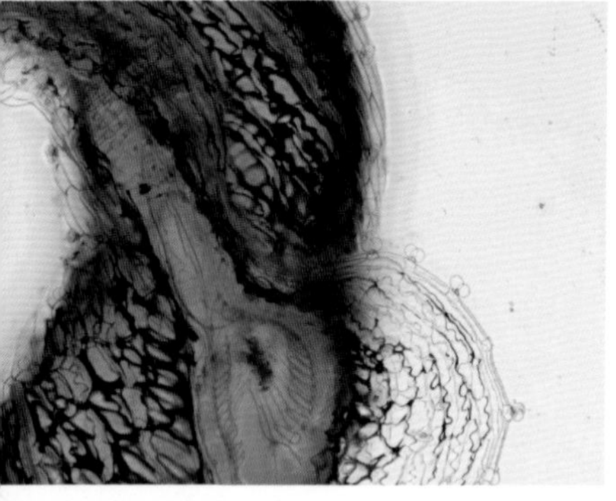

A microphoto of the bristly interior of a corkscrew trap, with prey.

Genlisea send up small, pretty flowers that look rather similar to those of some bladderworts. The flowers of the species are usually yellow or violet, although white forms are known as well. Several species are now in cultivation.

My favorite is *G. hispidula* from Africa, which has dark pink blooms. I have had plants of this species growing in the same large goblet in my hothouse for many years. Other easy species are *G. repens* and *G. pygmaea*, both yellow-flowered plants from South America, and *G. violacea*, a violet-flowered plant. Some *Genlisea* species produce flower stalks that have glandular sticky hairs that catch tiny insects, but it is as yet unknown whether this is simply defensive or the plants benefit from these prey. At least one species, *G. aurea*, has its surface photosynthetic leaves covered in a thick, clear mucilage for unknown reasons. Fernando Rivadavia discovered that another species, *G. pygmaea*, produces tubers to survive the annual dry season. Fernando also introduced into cultivation *G. uncinata* from Brazil, which has thick and astoundingly tall flower scapes, up to 3 feet (0.9 m) high.

If you grow your corkscrew plant in water, introduce daphnia to feed them. Otherwise, an occasional foliar mist of fertilizer on the rosette leaves will keep them happy.

Propagation is easily accomplished with leaf cuttings. Gently pluck a few leaves from the surface rosette, including the white base. Lay these flat on peat or sphagnum, keeping the soil very wet. Each leaf will grow into a new corkscrew plant.

The Devil's Claw

(IBICELLA LUTEA, PROBOSCIDEA LOUISIANICA, AND P. PARVIFLORA)

The jury is still out on whether these and related species are carnivores or subcarnivores, and I won't go into the convoluted taxonomical name changes these plants have historically gone through. Apparently, in 1916, an Italian botanist named Eva Mameli conducted experiments on *Ibicella lutea*, then known as *Martynia lutea*. She believed she proved its carnivorous nature by feeding it egg whites and insects, noting how the food dissolved and was

Ibicella lutea.

absorbed by the plant. Unfortunately, she published her work in an obscure university publication that no one read.

In 1993 her paper was rediscovered and the species reexamined. This and other similar species from the *Proboscidea* family of unicorn plants and devil's claws catch insects with sticky glands. Some, like *Ibicella lutea*, grow on desert fringes, placing them among the few varieties that grow in dry habitats.

Ibicella lutea grows in the Sonoran Desert of Mexico and has been naturalized in Southern California and Arizona. The plants are annuals, the seed germinating with the arrival of the summer thunderstorms. They grow rapidly in the summer heat, thick fleshy stems scrambling along the ground for a few feet (1.8 m). The leaves, up to several inches (15 cm) long, grow scattered along the stem. They are spade shaped with strong veins. In structure, the plants are vaguely similar to geraniums, some varieties of begonias, or melon plants. The two related species, *Proboscidea louisianica* and *P. parviflora*, grow in much of North America and appear quite similar; the primary differences are flower color.

The whole plant—stems, leaves, even the sepals of the flowers—is covered in tiny glandular hairs. Under a magnifying glass, these glands appear almost identical to the hairs on butterworts and rainbow plants. The plant glistens in sunlight and feels sticky to the touch. The plants catch numerous tiny insects the size of gnats and midges.

The devil's claw seedpod, with seed.

The flowers of *Ibicella lutea* are large and funnel shaped like a bell or trumpet, with ruffled petals. Rather handsome, the corolla tube is yellow-green on the outside and bright yellow inside with some red spots along the palate. The two *Proboscidea* species have similar flowers that can be shades of purple, pink, white, yellow, or orange.

The creepy-sounding common name, devil's claw, comes from the awesome seedpods the flowers produce at the end of the growing season. They look like some horrific insect jaws, black and spiny. Native Americans often sell them to tourists as curios, and they grow *P. parviflora* for food and textile materials. The fleshy immature seedpods are eaten cooked or pickled. The clawed seedpods are probably adapted to be dispersed by larger mammals such as coyote, antelope, or bighorn sheep, or now-extinct megafauna, the pods attaching themselves to their fur. The plants are banned in some countries, such as Australia, because grazing animals like cattle and horses can get the clawed seedpods impaled in their hooves or mouths, often causing a hideously slow death, the seeds later germinating around the bloated, rotting bodies.

Obsessed investigator Barry Rice, while studying naturalized devil's claws in California, discovered a few species of stilt bugs and other assassin-type insects harmlessly negotiating the sticky leaves. It's possible their relationship with the plants may serve a purpose similar to that of the capsid bugs on *Roridula*, feeding on trapped prey while exuding nutrients onto the plant through their feces.

Devil's claws are easy to grow if you can get the hard, black, 1/4-inch (6 mm) seed to germinate. The plant can be grown in most climates that have a long hot summer. They often do well as a regular garden plant, grown similarly to tomatoes or melons, and they are excellent as a control for whitefly. In a pot, use some houseplant soil mixed with plenty of sand, perlite, and lava rock, similar to a mix for cacti. Water heavily and allow the soil to freely drain and become slightly dry between waterings. Lots of sun and heat in summer are beneficial.

To treat the rock-hard seed, I can recommend several methods. Sow the seed about 1/2 inch (1.3 cm) below the soil surface and keep moist, with warm days. You can pour boiling water over them (but do not soak them in boiling water). Try cracking them slightly with a hammer, or soak them in water for a few days (a drop of SUPERthrive per cup [235 ml] of water might help).

The plants die after setting seed, which should be stored in the refrigerator.

Philcoxia

This utterly strange genus had its first species discovered in 1966, but it took more than three decades before the genus was established, containing three rare species from central Brazil: *P. bahiensis, P. goiasensis*, and *P. minensis.*

Philcoxia bahiensis, a new carnivorous plant from deserts of South America.

The plants grow in pure white quartz sand in a climate of near-desert conditions: extremely hot days, cooler nights, and very little rain. It is not clear whether they are perennial or annuals, returning from seed each year.

From a small swollen stem or taproot the minute "leaves" are disk shaped at the end of a filamentlike stalk. The leaves are underground, just below the sand surface. The top of the disk is covered in very tiny glands that secrete drops of glue, similar to much larger butterwort leaves. The prey appears to be nearly microscopic nematodes.

The most visual aspect of *Philcoxia* is the pretty flowers, pink or purple, emerging about 8 inches (20 cm) above the sand on zigzagging stalks. They appear so much like bladderwort flowers that the genus was at first thought to be related to *Utricularia*.

As of this writing the genus has not entered cultivation, but I suspect it eventually will, first at botanical gardens. I imagine the best environment for them would be a hot, dry cactus greenhouse, grown in pure quartz sand with very high light intensity.

Roridula

Roridula is a genus of two species from South Africa that are extremely rare. The plants look so similar to sundews that they were briefly included in that genus in the nineteenth century. Francis Lloyd, in his 1942 book *The Carnivorous Plants*, dismissed them from the true carnivores because they do not produce their own digestive juices. The species are now considered subcarnivores, like *Byblis liniflora*, and they rely on assassin or capsid bugs to digest their prey for them.

Both species are scrubby, branching plants up to 4 to 7 feet (1.2–2.1 m) tall. The thin, woody stems are crowned with a dense cluster of small, thin leaves no more than 3 to 4 inches (7.5–10 cm) long. The branching stems on old plants can have up to thirty growing points on *R. dentata* and a few hundred on *R. gorgonias*, which are probably decades old. *R. gorgonias* grows near the cooler, foggy coast and has long, thin, lance-like leaves. *R. dentata* looks similar, but the leaves are serrated like a large-toothed saw, and it grows further inland. The climate is Mediterranean, with chilly wet winters and dry summers, but the plants grow in permanently moist to wet gravelly, peaty soils. The inland population of *R. dentata* can have much warmer summers than the coastal *R. gorgonias*. Both species have pink, five-petaled flowers with bright yellow stamens.

The leaves are covered with stiff, glandular hairs, but unlike those of the sundews they have no power of movement. The glue is more resinous than the viscid glue of *Drosera*. So many insects are caught on the plant—even prey as large as bees—that a lure is suspected, but none has been found. Matthew Opel of the University of Connecticut reports that his cultivated *R. dentata* plants exude the sweet aroma of vanilla.

An assassin bug stalking prey on *Roridula dentata*.

As mentioned in the introduction, only recently has it been shown that assassin bugs do the digestion of prey for the plant. Hordes of these tiny bugs live on the plant, able to walk all over the glands without being caught. When prey is trapped, the bugs close in, stabbing their needlelike mouths into the panicking, struggling victims. Slowly they suck their juices dry, leaving a shriveled carcass. Later the assassin bugs secrete clear drops of nutritious feces onto the leaves, which are then absorbed by the plant like a foliar fertilizer.

Roridula gorgonias.

Roridula is still very rare in cultivation. After a couple of attempts to grow the plants in my greenhouse, only to have the plants die suddenly without explanation, I decided on a new soil mix and environment for them. Living in a Mediterranean climate similar to their native habitat, my plants are now thriving outdoors in coastal Northern California. I grow it almost exactly as I do *Drosophyllum*, the dewy pine.

I use large terra-cotta clay pots at least 12 inches (30.5 cm) wide. I cover the drainage hole with a handful of long-fibered sphagnum. The soil recipe itself is equal parts sphagnum peat moss, sand, perlite, pumice, and lava rock. The plants sit in shallow trays outdoors in full sun, and I keep the soil wet to moist by watering from overhead until some water drains into the tray. Just as the tray water evaporates, I again water from overhead. I never let them dry out; during our winter rains the plants can be quite wet for long periods, but the gravelly soil allows good drainage. They can take light frost; once a brief harder freeze of 25°F (-4°C) damaged the grow points of a couple of plants, but soon new shoots appeared from the stem.

The plants seem unaffected by brief heat waves to 100°F (38°C), but our summer nights always cool dramatically even after a hot day. Matthew Opel grows *R. dentata* in a well-ventilated cool greenhouse using a sphagnum peat moss and perlite soil, well drained and not sitting in water trays; Opel admits that continuously warm, stagnant air invites fungal infection.

I use Maxsea fertilizer foliarly, but even without assassin bugs, trapped insects often defecate, and it appears the leaves absorb these nutrients.

Roridula flowers will self-pollinate but produce few seeds; these look not unlike mouse droppings. I have had seed germinate without specialized treatment. However, since periodic brush fires cause massive seed germination in the wild, fire treatment similar to what I suggest for *Byblis gigantea* may help germinate stubborn seed. It is much easier to soak the seed in the liquid smoke used to flavor meats, available in grocery stores.

Triphyophyllum peltatum

This odd plant had been collected and preserved in herbariums a few times since the mid-1800s and was described and named in 1951, and although several botanists recognized its similarity to other known carnivores, it wasn't until 1979 that Sally Green and others made observations in the wild and at Kew Gardens that proved the plant is carnivorous. Strangely enough, the plant eats insects during only one stage in its life.

Triphyophyllum grows only in West Africa—Sierra Leone, Liberia, and the Ivory Coast—and is on the verge of extinction. Its name relates to the three types of leaves it produces during the course of its life.

A carnivorous leaf of *Triphyophyllum peltatum.*

In its juvenile stage the plant produces a woody stem up to 3 feet (0.9 m) high, topped with a crown of lance-shaped leaves 14 inches (35.5 cm) long.

Then the leaves become carnivorous. The midrib of the flat leaves extends into an upright, wiry leaf that looks and behaves much like a leaf of the dewy pine (*Drosophyllum*). These filiform leaves have stalked gluey glands that catch many crawling and flying prey. Flat, sessile glands secrete juices that digest the prey. These carnivorous leaves may be entirely threadlike or transitional, flat near the base and filiform at the tip. The plant probably becomes carnivorous to accomplish its next feat—a climbing, flowering stem—which no doubt requires a supplement of nutrients.

This climbing stem can reach an astounding 150 feet (45.7 m) in height, where it blooms in the sunlight at the top of the forest canopy. The leaves of the stem look surprisingly like *Nepenthes* leaves, lance shaped with a strong midrib. Here the midrib does not produce tendrils and pitchers, but instead two short hooks, which it uses to climb up the surrounding vegetation. At the top, small clusters of fragrant white flowers appear. These turn into large seed, 4 inches (10 cm) across, that look like bright red umbrellas. Detached, they float away in the wind.

Since the first edition of this book appeared, *Triphyophyllum* has finally entered cultivation in a few botanical gardens in Europe and Africa, and tissue-cultured plants have been created. It requires large pots of a peaty, gravelly, well-drained soil, deep shade during its younger life, and extremely high humidity and hot temperatures. Due to the enormous climbing vine, a very large stove house is required. In cultivation, flowers have been pollinated and produce seed. If the climbing vine is cut back, new shoots appear at the base and the plant returns to its carnivorous stage. Stem cuttings may be possible.

APPENDICES

I. Tissue Culture

Tissue culture is a fascinating, sometimes fairly scientific way of propagating carnivorous plants in test tubes, and many websites come up on an Internet search. The first that I can recommend is the online encyclopedia www.wikipedia.org. Search for "plant tissue culture" and detailed information comes up, from introductory to more complex scientific links. A second popular site is www.kitchenculturekit.com, which offers instructions and supplies for having fun with tissue-culturing plants at home.

II. The International Carnivorous Plant Society (ICPS)

If you have found this book interesting and wish to learn more about the world of carnivorous plants, the hub of all Internet links begins with the ICPS website at www.carnivorousplants.org.

Here you will find a vast amount of information, including databases offering thousands of photographs, cultivar registration, educational articles, conservation programs, forums, and archives. One of the most popular features is the society's web ring, where you will find many links to wholesale and retail vendors, as well as popular websites managed by individuals and local societies associated with the ICPS from all over the world.

If you join the society, you will receive its quarterly journal, the *Carnivorous Plant Newsletter*, in which the most up-to-date cultivation and scientific articles appear. I have been a member of and writer for *CPN* for several decades. Currently membership is only $35 a year.

III. Conservation

The devastation and disappearance of carnivorous plant habitats throughout the world is currently beyond alarming.

This tragedy became particularly personal for me a number of years ago, when I returned to the site where I had first discovered carnivorous plants as a child, as mentioned in the introduction of this book. Accompanying me on my return to the lakeside habitat in southern New Jersey was my young nephew, who wanted to see the place where his uncle found the strange plants that had changed his life so many years before.

What we found was a scene of numbing horror. The once-picturesque setting of southern white cedars, sphagnum moss, pitcher plants, and sundews was now a bulldozed clearing, scraped clean by the machines that sat idle that dark Sunday afternoon, beside a debris pile of trees and vegetation the size of a small mountain. A sign read, "Coming Soon! Lake View Homesites!" My nephew picked up the shriveled leaf of a *Sarracenia purpurea*. When he asked me if it was a pitcher plant, I told him yes, it was. Clear to the water's edge, we found no living plants at all.

Similar scenes have occurred countless times for nature lovers and carnivorous plant enthusiasts around the world, and with growing frequency. In the United States, more than 95 percent of the original carnivorous plant habitats along the southeastern coastal plain are gone. That includes virtually all of our *Sarracenia* species, as well as the Venus flytrap. In fact, at the current rate of destruction, according to Dr. Thomas Gibson of the University of Wisconsin, the Venus flytrap will be extinct in the wild in roughly the next century, surviving in only a few preserves. Dr. Donald Schnell, one of the founders of ICPS, is frequently asked by enthusiasts for lists of easily accessible roadside habitats so they may see *Sarracenia* in the wild. In 1980, he offered a list of twenty sites. By 1995, his list had been reduced to three. In southwestern Australia, most *Cephalotus* habitats are now golf courses and housing tracts. In Southeast Asia, *Nepenthes* are fast disappearing due to clear-cut deforesting, and as a result some mountain habitats are experiencing climate change, with longer and more frequent droughts. *Aldrovanda* is already gone in Japan and is vanishing in Europe. *Darlingtonia* bogs along the Oregon coast are being drained and filled for convenience stores and condominiums. The depressing list goes on and on.

There is hope. The ICPS has conservation programs that memberships help pay for and volunteers help to enforce; these include efforts such as clearing out scrub and trees from bogs where naturally occurring fires are discouraged, as well as doing controlled burns. They report when property goes for sale that is habitat for CPs, before it is purchased, drained, and cleared for the next strip mall. They monitor and report poaching on already protected property. They build eco-friendly boardwalks in preserves open to visitors to protect habitats.

Many other organizations work hard to protect and restore CP habitats, and as a result save many other species as well. In thirty countries including the United States, the Nature Conservatory (www.nature .org) buys and protects endangered habitats, or works with landowners to save threatened species. The Atlanta Botanical Garden (www.atlanta botanicalgarden.org) has programs for saving or restoring wetlands in the South. The North American Sarracenia Conservancy does critical preservation work (www.nasarracenia.org). Similar organizations around the world work toward similar goals. Government-owned state parks and preserves do likewise. Support them!

IV. Laws and Commerce

Did you know that if you are caught digging up Venus flytraps in North Carolina—or collecting their seed—you could face up to $50,000 in fines and/or one year in prison? Or that if you take a vacation in Borneo and dig up some roadside *Nepenthes*, then are caught trying to smuggle them into the United States, you could face two years in federal prison, plus $10,000 in fines, plus an equal amount in attorney fees? Plant lovers, be informed! Laws protecting threatened or endangered species are a serious matter and should not be taken lightly.

The laws regulating trade of some endangered carnivorous plants can be a headache-inducing confusion of bureaucratic red tape, but keep in mind, these laws are meant to help these species survive. Here I will offer a very brief review of a few of these laws and list some agency sources for information and permits so trade in these species can be done legally.

The Convention on International Trade of Endangered Species (CITES) came into existence in the early 1980s. Many countries participated in the development of these laws to protect endangered species from being collected from the wild populations of one country and then shipped to another—usually for commercial reasons. For example, it has been estimated that prior to CITES up to 250,000 Venus flytraps per year were being removed from the Carolinas and shipped overseas. With CITES, commercial producers who ship flytraps overseas must obtain permits proving that the plants were commercially propagated. These laws govern personal, noncommercial international trade as well.

CITES divides species into two sections: those in danger of extinction are listed under Appendix I and those that are threatened with possible future extinction are listed under Appendix II. Plants under Appendix I require permits from both the receiving country and the country of origin. Plants listed under Appendix II do not require an import permit, only an export permit from the country of origin.

Appendix I species are:

Nepenthes rajah and *N. khasiana*; *Sarracenia oreophila*, *rubra* ssp. *alabamensis*, and *rubra* ssp. *jonesii*.

Appendix II species are:

All *Byblis* species, *Dionaea muscipula, Darlingtonia californica*; all *Sarracenia* species not under Appendix I; all *Nepenthes* species not under Appendix I.

Appendix II species are exempt from CITES regulations if material being shipped internationally is seed, pollen, or tissue-cultured material in flasks. Commercial nurseries that export CITES-listed plants that are not in flask provide export permits, but you may require your own import permits. The nursery you are dealing with can often help with this; otherwise, go to www.cites.org to obtain legal documentation.

In addition to the CITES permits, if required, all plant material being imported to or exported from many countries is required to have phytosanitary certificates showing that the plant material is free of pests and diseases and is being shipped in artificial soil. Exporting nurseries usually provide such phyto certificates, or they can be obtained through local agricultural departments. Individual countries have their own laws pertaining to trade of endangered species domestically. In the United States most of these regulations fall under the Endangered Species Act (ESA). Currently four species are covered by the ESA: *Sarracenia oreophila, S. rubra* ssp. *alabamensis, S. rubra* ssp. *jonesii*, and *Pinguicula ionantha.*

These species cannot be removed from wild populations. It is also against the law to import or export these plants or to ship or trade them between states without permits.

V. Taxonomy (and Other Vexing Annoyances)

"I mean, doesn't it have a scientific name?"
"Yes, of course, but who can denounce it?"

—*Little Shop of Horrors*, 1960

Taxonomical Wars

The war between horticulturalists (those who grow plants) and taxonomists (those who name plants) has been a fact of botany for more than 150 years.

It was Carl Linnaeus in the early 1700s who invented the brilliant idea that all species of life be given two names in Latin, a genus name and a species name, so anyone would know which plant or animal was being discussed. This is known as binomial nomenclature. Later, subspecies, varieties, and forms were added, since very few forms of life could be identified so simply due to the great variability that occurs in nature.

The "war" arose because taxonomy is a very complicated and evolving science; changes occur almost continuously, and horticulturalists often sigh and shake their heads when someone publishes a new scientific paper arguing for changes in a particular plant's name. This occurs not only among carnivorous plants but also with palms, and orchids, and cacti, and nearly everything else people love to grow.

One important thing to remember is that these name changes are opinions and are certainly not set in stone. Some experts may argue against such changes, or offer a different interpretation. Eventually a consensus of opinion rules the day, but this can take years or decades to resolve.

Taxonomists are also frequently at war among themselves. This is often humorously referred to as the war between the "lumpers" and the "splitters." The former group tends to think that if it looks like a duck, walks like a duck, and quacks like a duck, it's a duck! The latter will point out differences in the feather coloration or webbed feet and argue that it's a different type of duck that should be given a new name.

There are many examples of these wars among those who study carnivorous plants. For instance, some will argue that the species *Sarracenia rubra* is a complex of five subspecies: ssp. *rubra*, ssp. *jonesii*, ssp. *alabamensis,* ssp. *wherryi*, and ssp. *gulfensis*. This is a lumper argument. However, some splitters have forcefully suggested in published scientific papers that a couple of these should be species in and of themselves, such as *Sarracenia jonesii* or *S. alabamensis*. Others have argued that *S. purpurea* ssp. *venosa* var. *burkei* should be a separate species called *S. rosea*. Usually these differences are based on flower structure as well as leaf design, but also genetics.

Who is right? Again, it's a matter of opinion. Ironically, you don't have to have a degree in botany to argue these issues. In fact, some of the best-known names in the carnivorous plant taxonomical wars have been medical doctors or school teachers or astronomers! You don't have to be a trained botanist, but you do have to be smart and opinionated and have relatively thick skin when others begin to criticize your arguments!

Personally, I tend to side with the lumpers. As a nurseryman, on hearing of yet another Latin name change, I just groan and sigh and change a plant's name tag . . . eventually!

Complex Crosses and Greges

A grex (greges is plural) is a group name for all of the offspring produced by the artificial crossing of two parents. Currently, the International Society for Horticultural Science (ISHS) allows the use of grex names only among orchids. This is a vexing annoyance for many CP growers and nurseries. Let's take a look at a hybridization example used in the American Pitcher Plants chapter:

First cross: *S. purpurea* × *flava*.
Second cross: *S.* (*purpurea* × *flava*) × *leucophylla*.
Third cross: *S.* [(*p* × *f*) × *l*] × *minor*.

The conundrum is obvious. In this hybrid example, it is already difficult to simply say the hybrid's name: "*Sarracenia purpurea* by *flava* by *leucophylla* by *minor*." If one indicated the subspecies, forms, and varieties of the plants involved, one could end up with: "*Sarracenia purpurea* subspecies *venosa* variety *burkei* by *flava* variety *ornata* by *leucophylla* variety *leucophylla* by *minor* variety *okefenokeensis*." If you think *that's* exhausting, imagine writing it on a plastic name tag! Even using code letters, you would have: S. [(pvb × fo) × ll] × mo. And further imagine that your cross has 189 seedling offspring—and you don't have an expensive label printer!

If CP growers were allowed to name greges, all of the seedlings of the example cross could be given one name, such as S. × 'Godzilla'. They could indicate the individual plants are a grex by adding the letters Gx.

There are already "rogue growers" doing this, and I am among them. On our website for California Carnivores, we list the names of greges and indicate what their parentage is, so customers can look them up. Be mindful that a written description of the grex would be impossible, for genetics would cause tremendous variations in the offspring. Keeping lists would be of tremendous help in the future, for when sanity eventually comes to the ISHS, and the International Carnivorous Plant Society (ICPS) is allowed to legally use the grex system, lists of greges can be transferred to the website of the ICPS, having already been created.

As of this writing, members of the ICPS are organizing a method of registering grex hybrids, but this website will most likely remain separate from the society, until permission is granted by the overseeing ISHS.

Registered Cultivars

When the first edition of *The Savage Garden* was released in 1998, much of the nomenclature of carnivorous plants was in disarray. One primary example was the issue of cultivars, or cultivated varieties. All carnivorous plant growers at the time believed that a cultivar should be a plant of such extraordinary quality that it should be registered as a cultivar and usually reproduced only vegetatively to ensure all propagated plants were an exact duplicate or clone of the original. Adrian Slack, in his 1986 *Insect-Eating Plants and How to Grow Them*, discussing cultivars of *Sarracenia* hybrids, wrote: "I hold that the very best of the seedlings of such carefully conceived crosses—say, one in 3,000—should be selected and named . . ."

In the first edition of this book, following this principle, I put actual published cultivars in single quotes. However, I also had innumerable plants to discuss that had "nicknames" or site location names. Thus, I described registered cultivars such as *Drosera filiformis* 'California Sunset', but also many other nicknamed plants, like *Drosera filiformis* ssp. *filiformis* "Florida Giant," which is simply a taller variety of the plant found in Florida. I put cultivars in single quotes, and all other nicknamed plants in double quotes.

Since that time there were great changes in the International Carnivorous Plant Society with regard to naming conventions. The plants I once described with a double-quoted nickname are now considered varieties in cultivation, and thus cultivars that should be officially registered as such. There has been an effort to officially register my nicknamed plants as cultivars, but since there are hundreds of them, it will take some time. In this edition of the book, all cultivars are now delineated by single quotes, regardless of whether they have been officially registered yet.

For more information on cultivars, go to the society's website at www.carnivorousplants.org.

VI. Other Resources

SCOTT BENNETT

Scott Bennet has been drawing and painting most of his life, and is recognized by his professional peers, art critics, and collectors to be making the highest level art in what can be called the major tradition of visual art. Among his many influences are Bonnard, Matisse, Cezanne, Olitski, and Marin. While his major output since 1990 has primarily been painting landscapes and still lifes, he has also quietly been making botanical paintings and drawings of carnivorous plants that he happens to grow. Many examples are in this book. For more information about Scott Bennett's work, visit these websites:

sbennett.neoimages.net

www.flickr.com/photos/scottbennettstudio

To purchase prints of his botanical works, visit this website:

www.scott-bennett.artistwebsites.com

BARRY RICE

Longtime editor of *CPN*, Barry has a website, www.sarracenia.com. Informative, funny, and opinionated, Barry not only discusses CPs in general but also recounts many of the field trips he takes with his wife Elizabeth Salvia. I totally enjoy their antics!

CH'IEN LEE

An American who moved to Borneo, Ch'ien has discovered many new *Nepenthes* and works to preserve Borneo's vast forests and wildlife. He also offers awesome eco-tours and photography through his website www.wildborneo.com.my. Take the trip of your life!

STEWART MCPHERSON

A profilic writer (to say the least!), Stewart has written an incredible series of stunningly colorful books on carnivorous plants. They were invaluable to me in revising *The Savage Garden*. His website is www.redfernnatural history.com.

ACKNOWLEDGMENTS

I am still deeply thankful to everyone I acknowledged in the first edition of *The Savage Garden*, some of whom I will thank here again for their kind support and enthusiasm.

But first let me offer my sincere appreciation to all readers and reviewers of the original edition—your overwhelmingly positive comments have been far beyond what I ever expected when I first sat down to write *The Savage Garden* in 1997. I thank you all!

Among the staff and editors of the International Carnivorous Plant Society and its journal the *Carnivorous Plant Newsletter*, both retired and current members are among my heroes: Joe Mazrimas, Dr. Don Schnell, Leo Song, Dr. Larry Mellichamp, Dr. Jan Schlauer, John Brittnacher, Cindy Slezak, and especially my thanks to Bob Ziemer and my good friend Barry Rice. Special thanks to Geoff Wong.

Those involved with the (San Francisco) Bay Area Carnivorous Plant Society are a fun and supportive group of friends, but a particular thank-you goes to Larry Logoteta, Stephen Davis, Doris Quick, Scott Hootman, and Chuck Powell, and a kiss for Judith Finn, who enjoys my sick humor.

While I thank those who helped with the often bewitching and hypnotic photographs and illustrations for this edition of *The Savage Garden* on a separate page, and am deeply grateful to the many growers of these weird plants mentioned throughout this book, special thanks go to Ch'ien Lee, Fernando Rivadavia, Andreas Fleischman, Robert Cantley, Greg Bourke, Tim Bailey, and Stewart McPherson.

It was Lau Hodges of the San Francisco Conservatory of Flowers who asked us to do the first Chomp! exhibit, using mostly plants from our nursery. Hundreds of thousands of people have seen these exhibits, and many botanical gardens followed her lead, even calling their shows "The Savage Garden." Lau is the hand model on the cover of this book.

A special thanks to the team at Ten Speed Press: publisher Aaron Wehner, editor Lisa Regul, copyeditor Kristi Hein, proofreader Karen Levy, and designer Chloe Rawlins.

This book would not exist if I hadn't had the insane idea back in 1989 to actually open a carnivorous plant nursery called California Carnivores, the first of its kind in the world. I am deeply in debt to the hundreds of newspapers, magazines, and television shows that have done stories on us—and all involved seemed to have had a lot of fun while doing so!

However, California Carnivores could never have survived without the support of our many customers and volunteers. In particular I would like to thank Patrick Hollingsworth; his wife, Louise; and their children, Scott and Grace. Tom Kahl's love of the plants and extraordinary talents are improving the nursery's infrastructure; his participation has been invaluable. Clark Barton, Zach Gibson, Dan Bouchard, Forrest Freund, Eda Zahl, Ken Collins, David Green, Sara Judkins, Dave Ostrem, Cindy and Mardi Lloyd, Beth Sampson, Clinton Regas, Gary Gipson, Mike Wang, Nani Fitzpatrick, and Axel Bostrom's enthusiastic work—well, I just can't thank you and the other volunteers enough! And special thanks to Mickey Urdea for reasons you well know!

My coworkers are angels sent from heaven! Norma McFaddan can pack mail order and pot up plants faster than a Venus flytrap catching a bug. Liz Brown, my longtime friend who computerized this manuscript—I love you. And I was blessed with a wonderful mother, Eleonore, who retired after fifteen years at the nursery—you are the best mom ever, Little Nell.

And finally—and most important—my business partner, Damon Collingsworth, who bought his first sundew from me at the age of twelve! Damon coordinated all of the photographs for this revised version of *The Savage Garden*, which includes photos of many of his own plants, taken at the nursery. Thank you for making California Carnivores the successful and happy place it continues to be!

ABOUT THE AUTHOR

PETER D'AMATO has been growing carnivorous plants for well over four decades. In 1989, he started the nursery California Carnivores, the first and only major grower of carnivorous plants that is open to the public with regular business hours. They are a tourist attraction in the wine country of Sonoma County, California, about an hour north of San Francisco. They house the largest collection of carnivorous plants on public display in the world, and they have a thriving mail-order business, website, and Facebook page with more fans than any similar site in existence. The nursery has been the subject of more than three hundred media articles and numerous television programs.

They can be reached at California Carnivores, 2833 Old Gravenstein Highway, Sebastopol, CA 95472. Phone: (707) 824-0433; www.californiacarnivores.com.

Some volunteers and staff with the author at California Carnivores. Left to right: Tom Kahl, Norma McFadden, Damon Collingsworth, and Peter D'Amato.

PHOTOGRAPHY AND ILLUSTRATION CREDITS

I would like to thank all of the photographers and illustrators for their generosity and hard work!

Antique botanical prints on pages ii, 10, 89, 125, 142, 170, and 287 from the collection of the author.

Illustrations on pages iv, 128, 164, 207, 210, 249, 271, 279, 292, 334, and 341 by Judith Finn.

Photos on pages 2, 3, 9, 16, 17, 26, 31, 32, 40 (center and right), 45, 47, 53, 68, 72-75, 78, 80-84, 94-105, 110-122, 127-139, 152, 162 (upper), 164, 165, 167, 171, 172, 173 (upper), 174-178, 182, 184, 185, 187 (lower), 188-192, 199-203, 205, 211-220, 240 (lower), 243, 245, 246, 248, 250, 252 (upper), 253, 254 (upper), 255-258, 260, 261, 269, 273 (upper and center), 274, 277, 278, 280 (upper), 281, 282, 293, 294, 299, 309, 311 (upper), 316, 318, 324 (upper), 326, 328 (lower), 329, 335, 342 (upper), 344, and 361 by Damon Collingsworth.

Photos on pages 8, 22, 24, 40 (left), 49, 69, 87, 108, 154-157, 162 (lower), 181 (upper), 193, 204, 232, 236, 251, 254 (lower), 308, 310, 325, and 339 by Jonathan Chester/Extreme Images, Inc.

Photo on page 13 by Peter D'Amato.

Photos on pages 29, 30, 267, 275 (upper), 279 (lower), 324 (lower), 337, and 342 (lower) by Patrick Hollingsworth.

Photo on page 50 (upper) by Rob Gardner.

Photo on page 50 (lower) by Ron Gagliardo.

Photo on page 56 by Jana Olson Drobinsky.

Photos on pages 58, 61, 92, 196–197, 224, 259, 286, 307, 312 (upper), 314 (upper), and 334 by Sharon Bergeron.

Paintings & line drawings on pages 76, 108, 153, 161, 194, 209, 234, 237, 239, 245, 248, and 255 by Scott Bennett.

Photos on pages 77 and 173 (lower) by Chuck Powell.

Photo on page 91 by Larry Logoteta.

Photo on page 107 by David Johnston.

Photo on page 123 by Graham Hunt.

Photos on pages 141, 146 (upper), 186, 228 (upper), 242, 321 (upper), and 349 by Stewart McPherson.

Photos on pages 146 (lower), 147, 296, 301 (lower), 304, 311 (lower), 313, 314 (lower), 315, 317, and 319 by Ch'ien C. Lee.

Photos on pages 150 and 301 (upper) by Glen Rankin.

Photos on pages 166, 187 (upper), 198, 272, and 273 (lower) by Chuck Ratzke.

Photos on pages 168 and 183 by Andreas Fleischmann.

Photo on page 180 by Greg Bourke.

Photos on pages 181 (lower), 240 (upper), 241, 244, 252 (lower), and 275 (lower) by Ron Parsons.

Photos on pages 226, 280 (lower), and 343 by Barry Rice.

Photo on page 228 (lower) by Isao Takai.

Photo on page 239 by Juerg Steiger.

Photos on pages 300 and 302 by Bill Baumgartl.

Photo on page 303 by Drew Martinez.

Photo on page 312 (lower) by Lee Maddox.

Photo on page 321 (lower) by Robert Cantley.

Photos on pages 327 and 328 (upper) by Geoff Mansell.

Photo on page 338 by Jason Ksepka.

Photo on page 341 by Geoff Wong.

Photos on pages 346 and 347 by Fernando Rivadavia.

INDEX

C

E

F

P

V

W

Y

Z